John Harper Long

Laboratory Manual of Elementary Chemical Physiology

And Urine Analysis

John Harper Long

Laboratory Manual of Elementary Chemical Physiology
And Urine Analysis

ISBN/EAN: 9783337140465

Printed in Europe, USA, Canada, Australia, Japan

Cover: Foto ©berggeist007 / pixelio.de

More available books at **www.hansebooks.com**

LABORATORY MANUAL

OF

ELEMENTARY CHEMICAL PHYSIOLOGY

AND

URINE ANALYSIS.

BY

JOHN H. LONG, M. S., Sc. D.,

PROFESSOR OF CHEMISTRY AND DIRECTOR OF THE CHEMICAL LABORATORIES IN THE SCHOOLS OF
MEDICINE AND PHARMACY OF NORTHWESTERN UNIVERSITY.

WITH NUMEROUS ILLUSTRATIONS.

CHICAGO:
E. H. COLEGROVE & CO.
1894.

Entered according to act of Congress, in the year 1894.
By JOHN H. LONG,
in the office of the Librarian of Congress, at Washington.

PREFACE.

This little book is an outgrowth of the course given to second year students in the Medical School of Northwestern University, and represents in its present form the practical laboratory work as now followed in the course of study required for graduation.

The time devoted to the study of chemistry and chemical physiology in our American medical colleges has been altogether too short, and only within very recent years, with a few notable exceptions, has any attempt been made to impart systematic knowledge in this direction. But now the views of medical teachers are undergoing a rapid change and among the scientific studies recognized as forming a rational groundwork for the study of medicine proper, physiological chemistry is justly regarded as of the first importance.

It is the belief of the Author that instruction in chemistry should be given through two years of the course of the medical student. While it may be urged that general chemistry forms no proper part of the curriculum of the medical school it is certainly true that no one of our schools can afford to abandon it until it is generally agreed to make it a requirement for admission. But this is a long way in the future. In his first year work, therefore, the student should be taught the elements of general chemistry both in the classroom and the laboratory, the laboratory work consisting of simple experiments followed by qualitative analysis. Such a course opens the way for the laboratory practice in chemical physiology in the second year, and this should be made practical and simple.

In the set of experiments offered to students the Author has aimed to select only such as illustrate some point of importance, and which, at the same time, can be easily performed. Complicated formulas and reactions which are so largely introduced in some of our recent books on medical chemistry, so called, the Author has thought proper to exclude as not in any degree essential.

This book consists of two parts, the first dealing with simple experiments in chemical physiology, while the second takes up the subject of urine analysis as properly following and partly illustrating the first. While intended primarily for medical students, it is believed that it contains not a little matter of interest to the medical practitioner as it discusses fully all the practical processes of urine analysis of value at the present time and in the chapters on chemical physiology presents several features more in detail than is customary in books of this class.

The Author can make no great claims of originality but would freely acknowledge his indebtedness to such standard works as the Physiologies of Landois and Stirling and Foster, the Physiological Chemistry of Hoppe-Seyler, the Urine Analysis of Neubauer, and Vogel and others. He would extend his thanks to Messrs. E. H. Sargent & Co., of Chicago, to Mr. A. Kruess, of Hamburg and to Messrs. Franz Schmidt and Haensch, of Berlin, for the use of cuts used in illustrations. Finally, he wishes to express his sense of obligation to his assistant, Mr. Chas. H. Miller and also to his colleague, Professor Mark Powers, for valuable aid rendered in the reading of proofs.

THE AUTHOR.

Chicago, Aug. 15, 1894.

TABLE OF CONTENTS.

PART I. ELEMENTARY CHEMICAL PHYSIOLOGY.

CHAPTER I. Introduction 1
CHAPTER II. The Carbohydrates. 19
CHAPTER III. The Fats...................... 59
CHAPTER IV. Proteids or Albumins................ 67
CHAPTER V. The Blood...................... 91
CHAPTER VI. Bone Constituents, Saliva, Gastric Juice, The Bile............................ 111
CHAPTER VII. Milk, Beef Extracts, Flour and Meal.. 121
CHAPTER VIII. Water and Air..................... 134
CHAPTER IX. Special Problems........... 143

PART II. URINE ANALYSIS.

CHAPTER X. Outline of Tests. Preliminary Tests... 177
CHAPTER XI. The Tests for Albumins...... 187
CHAPTER XII. The Tests for Sugar................ 213
CHAPTER XIII. The Coloring Matters in Urine. Biliary Acids........................... 242
CHAPTER XIV. Determination of Uric Acid........ 254
CHAPTER XV. Urea 263
CHAPTER XVI. The Determination of Phosphates and Chlorides......,....................... 282
CHAPTER XVII. The Sediment from Urine.......... 298
CHAPTER XVIII. Unorganized Sediments and Calculi. 321
APPENDIX. Test Solutions and Tables............... 343

Part I.

Elementary Chemical Physiology.

Chapter I.

INTRODUCTION.

CHEMICAL Physiology is in the main concerned with the study of certain organic compounds existing in the animal body and with the changes which occur in the various food stuffs from the time they enter the stomach until they leave the body, in the form of broken down excreta, by the lungs or through the kidneys. Beginning with the digestive changes in the stomach a series of complex reactions take place which we are able to follow only to a very limited extent. The formation of the important elements in the blood, bile, saliva, the gastric and pancreatic juices, the building up of the solid tissues of the body and the subsequent breaking down of the same, the final appearance of all the elements of these tissues or secretions in the urea, uric acid, phosphates and other products of the urine or in the carbon dioxide from the lungs involve a number of chemical reactions of the highest interest and importance. It is the province of chemical physiology to investigate as far as possible the conditions under which these changes take place, to study the constitution of the food products at the one end of the scale and of the excreta at the other and by means of laboratory experiments to bridge over the space between the two.

Chemical physiology attempts to duplicate in the test-tube or beaker the changes taking place in the body and to build up from the end products intermediate ones and those at the beginning. Most of the problems suggested in this direction are as yet far from solution. A study of the urine, expired air and gastric juice has shown the nature of these products under what may be considered normal conditions of the body. Further study has developed certain facts connecting variations in the urine, for instance, with variations in the condition of health in the individual. Study has shown that in health certain bodies are absent from the urine, while in certain diseases they are present, and from this fact there has been developed the very practical department of chemical physiology, known as urine analysis, as an aid to diagnosis.

Much of the experimental work connected with the investigation of problems in chemical physiology is of so difficult a character as to be quite beyond presentation in an elementary manual. Fortunately, however, other problems, and among them very important ones, are more easily approached, so that they can be worked out in the form of simple laboratory experiments.

The phenomena of salivary, gastric and pancreatic digestion, of coagulation of blood and of oxidation and reduction of its most important element, of the emulsification and absorption of fats and of diffusion and osmosis are certainly among the most important and interesting in the whole field of physiology. They have for the student a new meaning when he learns them, not alone from the printed page or the discourse of the lecturer, but also by his own laboratory experiments. Chemical physiology offers a vast field for the research of the specialist, but it offers also abundant simple illustrations for the beginner.

Chemical physiology is, in a sense, a branch of applied organic chemistry inasmuch as by far the larger number of compounds discussed are compounds of carbon such as are commonly termed organic. It was at one time supposed, in fact was maintained as one of the fundamental principles of chemical theory, that organic bodies, so-called, are produced only through the agency of the animal or vegetable cell, that is by what were termed vital forces, or by simple transformation of compounds so formed. The building up of a complex carbon compound from the elements of the earth or air was supposed to be a problem which could be solved only by the plant or animal as a function of its growth.

In 1828 Woehler showed that this view was a false one and that characteristic organic compounds can actually be built up in the chemist's laboratory. This was demonstrated first by the production of a body of the greatest interest to the physiologist, viz.: urea, which under the old view would certainly be one of the last to be considered within the possibility of artificial production.

The discovery of Woehler marks the beginning of modern scientific organic chemistry, but it proved of scarcely less importance in the development of physiology, as it paved the way for the preparation and study of a great number of compounds, occurring in the animal body, whose relations were imperfectly or not at all understood, or whose existence even in many cases, was not suspected.

Nature of Organic Compounds.

Organic compounds contain carbon as their characteristic element. Nearly all contain hydrogen. Oxygen is present in a very large number and nitrogen is present in many which are of importance to the physiologist. Sulphur is present in certain important organic bodies and in a still smaller number there are found phosphorus, chlorine and several other elements. Compounds of carbon are characterized by the fact that when heated alone or with certain oxidizing agents to a high temperature they are decomposed with liberation of carbon dioxide. The presence of hydrogen is shown under the same conditions by the formation of water. Nitrogen in organic compounds may be recognized in several ways but usually most conveniently by conversion into ammonia. Sulphur, phosphorus and chlorine are converted into sulphates, phosphates and chlorides which may be detected by the usual methods of qualitative analysis.

The proof of the presence of oxygen in an organic compound is often indirect. Some of these reactions may be illustrated by simple experiments, as follows:

For Carbon. Nearly all organic substances are decomposed when strongly heated. If heated in the air they give off inflammable vapors and often leave a residue rich in carbon which, at a high temperature, burns, forming carbon dioxide.

Solid organic substances when intimately mixed with a large excess of fine granular black oxide of copper and heated to a high temperature in a hard glass tube are decomposed and completely oxidized at the expense of the oxygen of the copper oxide. Carbon present is converted into carbon dioxide and hydrogen into water. If the combustion tube is furnished with a delivery tube which dips beneath the surface of clear lime water the characteristic reaction for carbonates is given. Organic liquids are characterized by being inflammable or by yielding inflammable vapors.

For Hydrogen. The presence of hydrogen may be shown by burning the substance in the air in such a manner as to condense the products of the combustion by a cold porcelain or metallic surface. The appearance of droplets of water is proof of the presence of hydrogen. In the oxidation by copper oxide, if the process is properly conducted, drops of water condense on the cooler parts of the combustion tube.

For Nitrogen. Nitrogen in most organic substances can be recognized by heating the latter with soda-lime to a high temperature. The nitrogen is given off as ammonia, which is recognized by the smell or reaction with litmus.

When a solid nitrogenous substance is heated in a narrow test-tube with some small fragments of clean metallic sodium, cyanide of the metal is formed. Nitrogenous liquids are mixed with dry sand and then with the sodium. The cyanide is dissolved out from the fused mass with distilled water in small amount, to the solution a drop of ferrous sulphate solution and a drop of ferric chloride solution are added and then hydrochloric acid to acid reaction. The appearance of the Prussian-blue color proves the cyanide and therefore the nitrogen. To recognize nitrogen in nitro-compounds a mixture of soda-lime and sodium thiosulphate may be used.

For Sulphur. When organic compounds containing sulphur are fused with a mixture of potassium nitrate and sodium carbonate a sulphate is formed which may be recognized in the dissolved mass by the usual tests.

Organic sulphur compounds, albumin for instance, give up their sulphur as sulphide when heated in a glass tube with metallic sodium. The sulphide can be detected in the usual way.

Classification of the Physiologically Important Organic Bodies.

From a strictly scientific point of view organic bodies should be grouped here as elsewhere, that is as products derived from fundamental hydrocarbons. But for many reasons such a classification would be inconvenient or even misleading for our purpose.

For the physiologist it is in most cases sufficient to consider bodies as belonging to, or related to members of, three important groups of natural substances, viz.: the carbohydrates, the fats and the albuminoids.

Certain compounds appear properly as decomposition products of members of one or the other of these groups.

By far the greater number of organic substances considered by the physiologist are bodies of natural origin, or, at any rate, difficult of synthesis. While several of the carbohydrates have been produced artificially, and fats have likewise been built up, no albuminoid has yet been made by an artificial process. The plant alone seems to have the power of building up proteid substances. In the animal body these are converted into new forms or digestion products, but cannot be created outright. Carbohydrates are produced also, in the main, in the vegetable cell, but fats can be built up in the animal body from substances not fat.

The animal, however, has not the power of producing any organic substance directly from the elements.

In the following, attention will be paid primarily to groups and derivatives as outlined.

General Methods of Experimentation.

It is assumed that the student beginning these experiments is reasonably familiar with elementary general chemistry and qualitative analysis. Most of the work following is of a qualitative nature, involving the use of very simple methods only, and such apparatus as the student has already used in other work.

A few volumetric quantitative processes are introduced, the details of which will be explained in the proper conection. The simple, general principles common to all volumetric methods may, however, be outlined here.

Quantitative determinations are made by *gravimetric* or *volumetric* methods. In the first the substance sought, or some compound bearing a definite and known relation to it, is precipitated, usually collected and weighed on a filter or in a crucible or other small receptacle. In volumetric methods the procedure is different. Advantage is taken of the fact that many reactions in solution are completed in some definite manner which permits the end-point to be accurately observed. For instance, if to some dilute sulphuric acid in a beaker, to which a little litmus or phenolphthalein had been added, a dilute solution of potassium hydroxide be next added a sudden change of color appears as soon as enough alkali is present to neutralize the acid and leave a trace in excess.

An alkali solution made blue with litmus becomes as suddenly red on addition of an acid.

If a solution of salt is added to a solution of silver nitrate, drop by drop, a precipitate appears. If the silver solution is contained in a bottle which is violently shaken after each addition of salt, the precipitate formed soon settles out so that it is possible to observe the formation of a new precipitate as fresh drops of salt are added. By working carefully the point can be determined where the addition of a drop of salt solution ceases to produce a precipitate in the silver bottle, and this indicates the end of the reaction.

Again, suppose we have a hot dilute solution of oxalic acid containing a small amount of sulphuric acid. To this is added, drop by drop, some solution of potassium permanganate, the deep color of which is very characteristic. As the drops of permanganate touch the acid solution the color disappears almost instantly, because of decomposition of the compounds, (oxidation of the oxalic acid). The instant the last trace of oxalic acid has been decomposed the solution is almost colorless, but the addition of a drop more of permanganate is sufficient to impart a pink color readily seen. In this case the end of the reaction is indicated by a sudden change of color.

We determine in these experiments the relation between acid and alkali, between silver and salt, between oxalic acid and permanganate, in solution and in the same vessel in which we began the test, and besides in a very short time.

It is evident that if we know the *strength*, grams per liter, of the potassium hydroxide, of the salt solution and of the permanganate we *determine* the amount of sulphuric acid in the first, of silver in the second and of oxalic acid in the last of the above experiments by noting simply the volumes of the alkali, salt and permanganate solutions needed to give the characteristic end reaction. This follows because we know that the several compounds react on each other according to the following equations:

$$2 KOH + H_2SO_4 = K_2SO_4 + 2 HOH$$
$$112 \quad + \quad 98 \quad = \quad 174 \quad + \quad 36$$

$$NaCl + AgNO_3 = AgCl + NaNO_3$$
$$58.4 \quad + \quad 169 \quad = \quad 142.4 \quad + \quad 85$$

$$2 KMnO_4 + 5 C_2H_2O_4 + 3 H_2SO_4 =$$
$$316 \quad + \quad 450 \quad + \quad 294 \quad =$$
$$2 MnSO_4 + K_2SO_4 + 8 H_2O + 10 CO_2$$
$$302 \quad + 174 \quad + \quad 144 \quad + \quad 440$$

112 Gm. of potassium hydroxide neutralize 98 Gm. of sulphuric acid; 58.4 Gm. of sodium chloride precipitate 169 Gm. of silver nitrate and 316 Gm. of potassium permanganate oxidize 450 Gm. of anhydrous oxalic acid.

The proportions hold good if we substitute here milligrams, grains, ounces or pounds for grams throughout.

If now we prepare solutions containing in one liter 112 Gm. of KOH, 58.4 Gm. of NaCl and 3.16 Gm. of $KMnO_4$

one cubic centimeter of these solutions will react with 98 milligrams of H_2SO_4, 169 milligrams of $AgNO_3$ and 4.5 milligrams of $C_2H_2O_4$.

Solutions prepared in this manner are called *standard* or *volumetric* solutions, because a definite volume of each reacts with a certain weight of something to be measured. A *normal* solution is usually defined as one which contains a weight in grams equivalent to the molecular weight of the substance in question, dissolved in one liter. Thus, 58.4 Gm. of NaCl dissolved in one liter would yield a normal solution. A tenth-normal ($\frac{N}{10}$) solution of salt would contain 5.84 Gm. in the liter. A twentieth-normal ($\frac{N}{20}$) solution would contain 2.92 Gm. of salt in a liter. A normal solution of sulphuric acid according to this definition would be one with 98 Gm. of H_2SO_4 in the liter. But many chemists take 49 Gm. of H_2SO_4 (half the molecular weight) as constituting here a normal solution. One cubic centimeter of this would exactly neutralize one cubic centimeter of normal potassium hydroxide (always taken as containing 56 Gm. to the liter) and would be equivalent in acidity to one cubic centimeter of a normal hydrochloric acid with 36.4 Gm. to the liter. From this standpoint a normal acid solution is one which furnishes in each cubic centimeter as many replaceable hydrogen atoms as are furnished by normal hydrochloric acid with 36.4 Gm. to the liter.

Many standard solutions are made empirically of such a strength that one cubic centimeter will react with some conveniently small whole number of milligrams of the substance to be measured. A well-known test for sugars depends on the reducing action of these compounds on alkaline copper solutions. Based on this behavior we have a volumetric process in which the standard solution (Fehling's solution) is made of such a strength that one cubic centimeter is completely reduced by 5 milligrams of dextrose. This important test will be explained in detail later. Another volumetric solution of frequent use in the physiological laboratory is the mercuric nitrate solution for measurement of urea. Under certain conditions of dilution urea and mercuric nitrate react with each other in a perfectly definite manner, the urea falling as a precipitate with the mercury. The end of the reaction is easily determined and the solution of mercuric nitrate is made of such a strength that one cubic centimeter precipitates exactly 10 milligrams of urea.

In volumetric analysis *indicators* are very frequently used to show the end of the reaction. These are usually substances which give a characteristic color change as soon as the principal reaction is completed. Litmus and phenol-phthalein are indicators which show change of reaction from acid to alkaline, or the reverse, and are added, in small quantity, to the solution whose strength is to be measured by a standard acid or alkali solution.

If to a salt solution whose strength is to be found by silver nitrate, a few drops of neutral potassium chromate solution be added, the chlorine is completely precipitated first, on adding the silver, and then deep red silver chromate. The appearance, therefore, of a faint shade of red in making the test *indicates* the end of the principal reaction.

In some cases the indicator is not added directly to the solution to be measured, but is put in the form of drops on a glass or porcelain plate; and with these, from time to time, a drop of the solution to which the standard is being added, is mixed. When enough of the standard has been used a color reaction usually appears in the mixed drops. The end of the reaction in measuring urea by mercuric nitrate or phosphates by uranium solutions is shown in this manner.

Measuring Apparatus. In volumetric analysis much naturally depends on the accuracy and convenience of the measuring apparatus employed. This apparatus consists of *flasks, mixing cylinders, open cylinders, pipettes* and *burettes.* The usual forms of these are shown in the accompanying figures. See next page.

Flasks are employed of the following dimensions : 1,000 Cc. (one liter), 500 Cc., 250 Cc., 100 Cc. and occasionally others. They are made to *contain* these volumes, and when correctly graduated *deliver* slightly less of all liquids except mercury.

Mixing Cylinders are usually employed of the dimensions given for flasks, but cannot be used where the greatest accuracy is required.

Open Cylinders are used for measuring volumes of liquids approximately, and are made to contain 500 Cc., 250 Cc., 100 Cc., 50 Cc., 25 Cc., 10 Cc. and 5 Cc., usually.

Pipettes are made to accurately *deliver* small volumes of liquids, and those most frequently used have these dimensions : 100 Cc., 50 Cc.. 25 Cc., 10 Cc. and 5 Cc. Pipettes are sometimes graduated to resemble burettes.

Burettes are made to *deliver* accurately small volumes of liquid down to one-tenth or one-twentieth of a cubic centimeter. They usually hold 50 Cc., sometimes 25 Cc. or 100 Cc. They are frequently closed below with a ground glass stop-cock. Sometimes with a rubber tube compressed by means of a brass clamp, or filled by a little glass ball inside. The ball is placed between the end of the burette and the fine delivery tip, and when the rubber tube is squeezed so as to form a channel on one side, the liquid is allowed to pass.

These different measuring vessels are readily found in the market graduated with accuracy sufficient for all prac-

12   *ELEMENTARY CHEMICAL PHYSIOLOGY.*

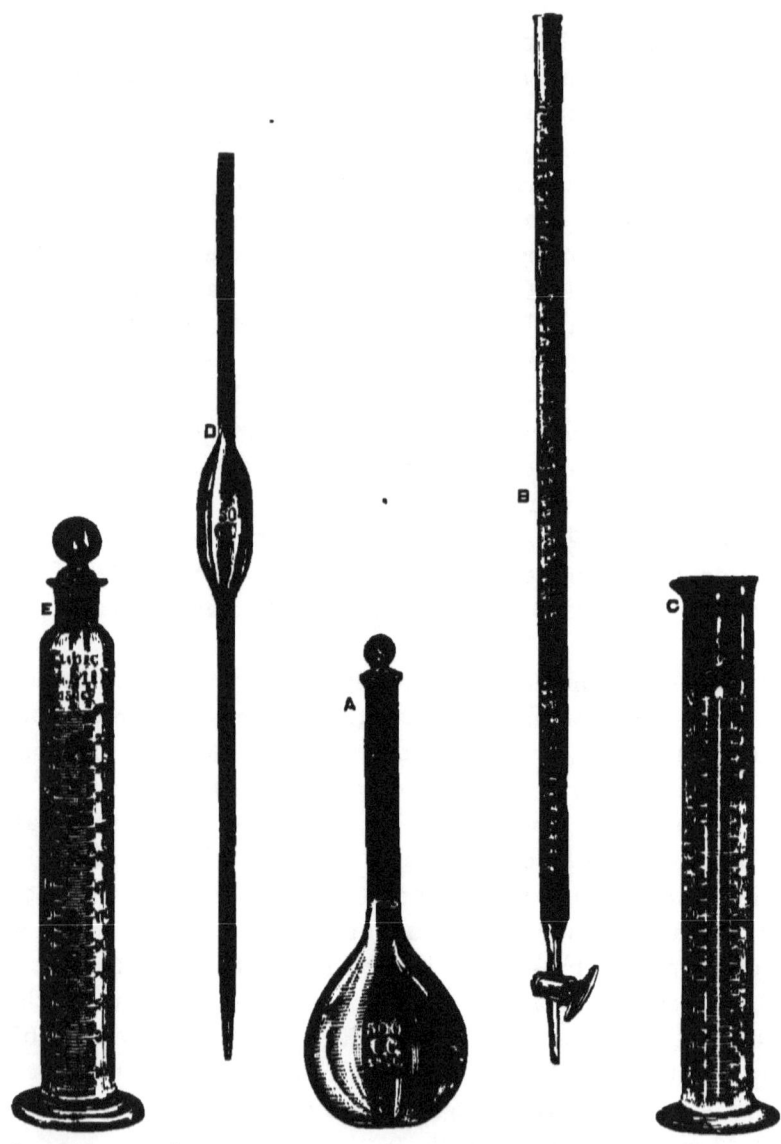

A represents a flask; B, a burette; C, an open cylinder; D, a pipette and E a glass-stoppered mixing cylinder.

FIG. 1.

tical purposes.* If the user desires to further test their accuracy, this may readily be done by weighing the amount of distilled water at some standard temperature, they hold or deliver.

Thermometers. While in ordinary qualitative analysis the reactions are carred out with little regard to definite temperatures, the room temperature or that of boiling water being sufficiently close for practical purposes, in chemical physiology, for a wide range of investigations, the temperature must be maintained within rather narrow limits. This is necessary because many chemico-physiological changes are brought about by the activity of living ferments or of enzymes which are destroyed by high temperature, and are either destroyed or become inert at low temperature. A certain class of phenomena are best observed at a temperature near that of the human body, or a little above, that is about 40^b C., while for other reactions a temperature of $65°-70°$ C. is found best, while a temperature of $75°-80°$ C. would be destructive.

For the ordinary uses of the chemical physiological laboratory a thermometer graduated in single degrees with a range $-6°$ C. to $110°$ C. is sufficient. Such thermometers can be purchased from chemical dealers and are usually made with accuracy enough for practical purposes. But this is not always the case, and the user should test his instrument to ascertain its approximate error.

First, the error at the zero point should be found by immersing it in *clean fresh* snow which has been kept an hour or more in a room with a temperature a little above the freezing point, $0°$ C. This will insure that the snow has the temperature of zero exactly if it has a chance to drain. Snow in this condition packs easily when pressed. The

---

*Graduated glassware and other apparatus described in this book may be obtained from Messrs. E. H. Sargent & Co., 106 and 108 Wabash Ave., Chicago.

mercury of the thermometer left in about half an hour sinks and finally becomes stationary. If the graduation is correct the top of the column should be at the 0° mark. Instead of employing snow, which is seldom found clean enough in cities, and the use of which is limited to the winter season, a better plan is to freeze some pure distilled water in a test-tube by surrounding it with a mixture of crushed ice and salt. When the water begins to freeze it is shaken to produce ice in fine particles. The bulb of the thermometer is now immersed in this loose ice, which, as long as any water is present, has accurately the 0° temperature, and stirred about. The level of the top of the mercury column soon becomes constant and can be read off. In many of the cheaper thermometers this level is somewhat above the zero mark.

The other fixed point on the scale is determined by suspending the instrument in the steam of water boiling in a tall metallic vessel. The bulb must not dip in the boiling water, and the top of the mercury column must be a trifle above the neck of the vessel.

The boiling point is dependent on the atmospheric pressure and is by definition 100° C., under a pressure of 760 Mm. For temperatures near 100° the boiling point is lowered 1° by a depression of 27 Mm. in the barometric column, from which datum a correction can be made.

The error of the thermometer at the two fixed points having been determined it remains to determine the errors in the intermediate parts of the scale. The readings at the fixed points may be correct and yet wrong at points between, because of lack of uniformity in the diameter of the capillary tube. To determine the errors accurately between the two fixed points is a matter of some practical difficulty, and will not be explained here in detail. An idea of the amount of the variations in the section of the capillary sufficient for our purpose can be obtained by shaking off a short thread of mercury from the column and allowing it to flow from one point to another, noting its length as measured by degrees on the scale. As the volume of the separated thread is constant a variation in the bore of the capillary is indicated by the lengthening or shortening of the thread as it moves through the tube. Large errors can be detected in this way; small ones we are not concerned with here.

Thermometers are graduated with the centigrade or Fahrenheit scale in this country. For all scientific measurements the former alone is used. The conversion of temperatures on one scale to temperatures on the other is a simple matter and can be done by these formulas. Centigrade temperatures to Fahrenheit:

$$F° = 32° + \tfrac{9}{5} C°$$

Fahrenheit temperatures to centigrade:

$$C° = \tfrac{5}{9}(F° - 32°)$$

Separation by Dialysis. For the separation of certain classes of compounds the methods of dialysis are very frequently employed. Many substances in solution possess the property of passing through animal membranes, while other bodies, likewise soluble, do not. Sugar and salt are types of the first class, while glue and albumin are types of the second. The phenomenon depends ultimately on a property common to all bodies in solution, the property of diffusion, which may be illustrated in this way.

Suppose we partly fill a glass jar with a strong solution of copper sulphate and then bring over this, with as little disturbance as possible, a layer of pure water and leave the jar so charged, several days in a perfectly quiet place at a uniform temperature. It will soon be observed that the blue color of the lower layer begins to diffuse upward, notwithstanding the fact that the copper sulphate solution is specifically heavier than water. After a long time the two layers become as one, molecules of water passing down to take the place of the copper sulphate molecules which ascended.

Something very similar takes place when a layer of water is brought over a layer of solution of salts, acids, alkalies or other substances. Here the change cannot be followed by the eye, however, but must be detected in some other manner. Solid substances in solution show the same tendency to distribute themselves through all available space that is characteristic of gases. The phenomena are the same with miscible liquids, e. g., alcohol over water. We speak, therefore, of diffusion of solids, liquids or gases.

This fundamentally simple phenomenon is greatly modified if we separate the substances by an animal membrane, as a piece of bladder. The rate of diffusion of many substances is retarded, while that of others is made so slow as to practically prevent passage. Graham, who investigated this subject carefully, divided substances roughly into *crystalloids* and *colloids*, crystalloids being those bodies which in solution diffuse through animal membranes, and colloids (glue-like substances) those which cannot pass through. It seems to hold true that the failure of bodies to diffuse through membranes depends on the size of their molecules, as the molecules of all those substances called colloids are known to be large, in some cases enormously large.

In physiological chemical analysis these principles are applied in what was termed above *dialysis*, or separation by diffusion through membranes.

This can be carried out by very simple appliances. A dialyser is easily made by stretching a piece of wet parchment over a hoop and tying it down close all around. This makes a shallow vessel into which the substance to be investigated is poured. The dialyser is then floated on water in a larger vessel. The water is renewed occasionally and so hastens the passage of diffusible substance through the parchment. As an illustration a solution of albumin containing salt can be purified in this manner. The salt diffuses while the albumin fails to.

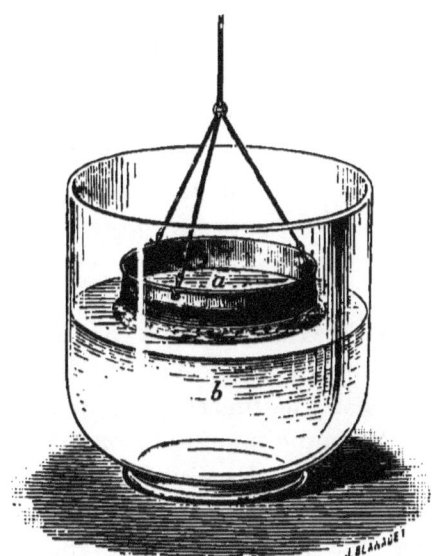

FIG. 2.

Instead of using the common hoop dialyser a parchment tube is now frequently employed. The tube is bent, making a narrow double sack into which the liquid to be dialysed is poured. The filled sack is then suspended in pure water which is renewed as before.

The cut above, Fig. 2, shows the manner of using the first form of dialyser.

Determination of Specific Gravity. By the specific gravity or density, or specific weight, of a substance we understand the ratio of the weight of a given volume of the substance to the weight of the same volume of some other substance considered as a standard. To make our definition exact we must suppose the volumes taken at some normal or standard temperature. The standard substance to which specific gravities of solids and liquids are referred is water and usually at a temperature of 4° C.

The temperature of the substance compared with water must be accurately known but may be 4°, 15°, 20°, 25° C., or in fact any temperature, but usually one of these. When we say the specific gravity of a liquid is 1.0154 we mean that a volume of it at some definite temperature is that much heavier than the same volume of water at 4° C. If taken at 20° C. we can express the true specific gravity in this manner.

$$\text{Sp. gr.}^{20}_{4} = 1.0154.$$

Simple relations exist between the densities of solutions and the amounts of dissolved substance, and the determination of density is, therefore, usually made in order to find amount of a body in solution or to detect variations in amounts of substances dissolved. The progress of fermentation in a sugar solution is readily followed in this manner, and the amount of solid substance in urine is approximately found by making a determination of its density.

Special application of these facts will appear later and in the appendix general tables of specific gravities of solutions will be given with methods of determination.

Chapter II.

CARBOHYDRATES.

The name *carbohydrate*, as commonly used, is given to a peculiar group of bodies containing carbon, hydrogen and oxygen combined in certain proportions. In this group are included the sugars, the starches, the gums and some allied substances. The molecules of the compounds in this group contain, ordinarily, six, or some multiple of six, atoms of carbon with twice as much hydrogen as oxygen and usually enough to yield five or more molecules of water. In a strictly scientific classification certain molecules with five or even four atoms of carbon should probably be included among the carbohydrates.

The carbohydrates are formed mainly in the vegetable kingdom and are, almost without exception, important food products. A few occur in the animal kingdom, but they are in this case derived from similar substances produced by plants.

In the experimental study of these bodies it is best to begin with the starches, which are widely distributed in nature and important in the highest degree.

Starch.

Ex. 1. Prepare starch from the common potato as follows: Grate a potato to a pulp by means of an ordinary tin grater, mix the pulp with water and squeeze it through a piece of coarse unbleached muslin. Moisten the pulp again and repeat this operation several times, collecting the strained liquids in a large beaker. Allow the mixture to settle a half hour or longer and pour off the water, which contains some soluble albuminous substances, some cellular floating matter, but very little starch. Most of this will be found in the bottom of the beaker. Add some fresh water, stir up and allow to settle. Now pour the water off again and repeat these operations until the starch appears perfectly clean and white. Transfer this starch to a clean shallow dish and allow what is not intended for immediate use to dry spontaneously in an atmosphere free from dust. The dried product will consist of minute glistening particles resembling small crystals.

Corn starch has been made from green corn, by a process quite similar to the above. Ordinary dry corn is used in the United States for the production of starch on the large scale, and to aid in separating the starch from the accompanying materials, it is customary to add a little acid or alkali to the water. Sometimes the meal is allowed to stand with water until a slight fermentation begins, which has the effect of disintegrating the cell walls holding the starch. Potato starch is the common starch of continental Europe.

Starch consists of minute microscopic granules varying in size and shape with the root or seed employed. The granules of wheat starch are regular in outline, but vary greatly in size, the smallest having a diameter of less than .005 Mm., while the largest may have a diameter of .04 Mm. Corn starch granules have a mean diameter of about .02 Mm., while those from the potato are often .06 Mm., or more in diameter. See Figs. 3 and 4.

Ex. 2. Examine starch from several sources under the microscope, employing a power of about 300 diameters. Clean a glass slide *thoroughly*, place in the center of it a small drop of water, and stir into this by means of a needle, or glass rod, a minute quantity of starch. Now drop on a thin, carefully cleaned cover glass, and in such a manner as to exclude air bubbles, and place under the microscope for observation.

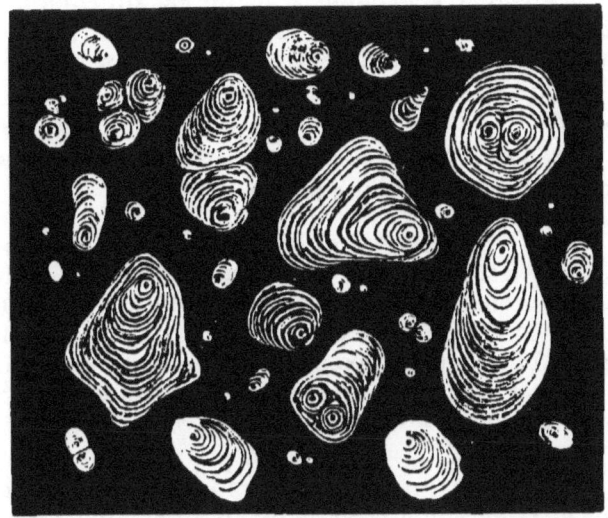

FIG. 3.
Potato Starch.   300 diameters

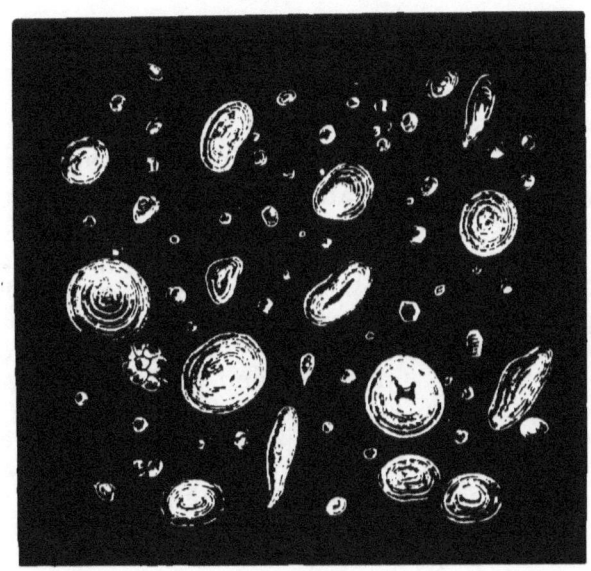

FIG. 4.
Wheat Starch.   300 diameters.

Ex. 3. Repeat Experiment 2, using an aqueous solution of iodine instead of water. The starch granules will now appear blue. For the detection of starch in mixtures the use of iodine is often indispensable.

Starch can be recognized by a number of chemical tests, the best of which are the following:

Ex. 4. Boil a small amount of starch with water so as to make a thin paste. Allow this to cool, and add a few drops of an aqueous, or dilute alcoholic solution of iodine. A deep blue color is formed which disappears on boiling the mixture. This test is exceedingly delicate and characteristic, and serves for the detection of minute traces of iodine as well as starch.

Ex. 5. Starch is insoluble in cold water, which can be shown by stirring some with water in a beaker, allowing to settle, and pouring the liquid through a paper filter. The filtrate tested with the iodine solution does not give a blue color.

Action of Acids on Starch.

When boiled with dilute acids starch is converted into soluble compounds. The nature of these compounds depends on the acid used and on the duration of the heating. Two experiments will illustrate these points.

Ex. 6. Make a paste by boiling about a gram of starch with 100 Cc. of water in a glass flask. Add ten drops of dilute sulphuric acid (1:5) and boil five to ten minutes. Now allow the liquid to cool, remove 5 Cc. with a pipette, dilute this to 25 Cc. with water and add a few drops of iodine solution; a blue violet color results, showing that starch or a starch-like substance is still present. The remainder of the acid liquid in the flask is next boiled steadily for one hour, a little water being added from time to time to replace that lost by evaporation. At the end of an hour remove 5 Cc., dilute and test with iodine solution as before. The characteristic starch reaction is now absent, while the liquid has become thin and transparent. Save this for experiments below.

Ex. 7. Add 5 Cc. of strong sulphuric acid to a gram of starch in a flask holding about 200 Cc. Heat to the boiling point and observe that a black mass is soon produced. By prolonged heating this is further decomposed, while fumes of sulphurous oxide escape, leaving finally a colorless liquid. This experiment must be performed in a fume chamber.

Ex. 8. Add 5 Cc. of strong nitric acid to one gram of starch in a flask holding 200 to 300 Cc., place this on a sand-bath in a fume chamber and apply heat. After a time copious red fumes are given off. Remove the lamp and allow the reaction to continue until the fumes cease to be evolved. Finally, transfer the liquid to a porcelain dish and evaporate to a small volume. On cooling, a crystalline residue remains which consists mainly of oxalic acid.

These experiments show in a striking manner the behavior of starch with acids. The action of dilute acids in general resembles that of sulphuric as illustrated in Ex. 6. This reaction is of great practical importance and should be carefully observed by the student. Therefore, with the remainder of the liquid left in Ex. 6 let the following experiment be tried:

Ex. 9. Neutralize the free sulphuric acid by addition of a slight excess of chalk or fine marble dust, heat gently to complete the reaction. Then filter and evaporate the filtrate nearly to dryness on a water-bath. Allow to cool and notice that the residue has a sweet taste. It is, in fact, dextrose and the experiment illustrates the method of manufacture on the large scale. Test it by dissolving a little in water and adding a few drops of *Fehling's solution*, (see appendix). On boiling, a yellowish precipitate appears which becomes bright yellow and finally red. Other tests will be given below under the sugars.

Much of the value of starch depends on the readiness with which it can be converted into sugar. In the animal body this reaction takes place normally and is brought about in two general ways, by the action of saliva and also by the pancreatic juice. Starch, being insoluble in water and very dilute acids, cannot be directly absorbed from the alimentary tract. It must first be brought into soluble form or converted into a soluble compound, and this is just what takes place in digestion. What are commonly termed digestive processes, are carried on in the body through the action of peculiar complicated chemical compounds called *soluble ferments* or *enzymes*. These are unorganized substances and not living cells as in the case of yeast, which may be taken as a type of what are called *organized ferments*.

The saliva contains an enzyme known as *ptyalin*, which has the power of converting starch into malt sugar or maltose. A similar active principle is found in malted grain, especially in common malt prepared from barley. The bitter almond contains also an enzyme, called *emulsin*. The behavior of these bodies will be shown by a few experiments below.

In the pancreas is found a soluble ferment which rapidly converts starch into sugar. The reaction is, in its main features, similar to that produced by saliva, but goes somewhat further. The ptyalin of saliva has been shown to yield maltose as the important product, even after protracted action. But in the case of the pancreatic ferment, while maltose is at first formed, dextrose very soon appears and may possibly be looked upon as a normal end product. The action on starch, begun by the saliva, is continued by the amylolytic enzyme of the pancreatic juice and completed, as comparatively recent investigations seem to show, by the secretion of the small intestine. This secretion has only a slight action on starch itself, but rapidly converts maltose into dextrose, which is a point of the highest importance, since maltose seems to be only imperfectly assimilable, while of dextrose the reverse is true.

ACTION OF SALIVA ON STARCH.

Experimentally the digestive behavior of saliva can be shown without difficulty.

Ex. 10. After washing out the mouth thoroughly with water, chew a piece of rubber or other neutral insoluble substance to stimulate the secretion of saliva. Collect this in a beaker and continue until about 25 Cc. has been secured. Dilute the more or less turbid liquid with an equal volume of water and allow to stand a short time, and then filter through a 7 or 8 Cm. filter into a small clean flask.

Ex. 11. Make a thin starch paste, about a gram to 200 Cc. of water, and observe that it does not respond to the Fehling sugar test referred to above. Mix 10 Cc. of this paste with 5 Cc. of the filtered saliva and warm to a temperature not above 40° C. for about fifteen minutes. At the end of the time apply the sugar test again. A yellow or red precipitate will appear now, showing that the starch has been converted, in part at least, into sugar.

The saliva alone fails to reduce the copper solution, as should be shown by trial.

Ex. 12. Pour about 5 Cc. of the clear saliva in a test-tube and boil a few minutes; add the starch paste and allow to stand as in the above experiment. On testing with the copper solution no sugar will be found, showing that heat destroys the activity of the ferment.

The digesting power of the saliva is destroyed also, by the addition of a small amount of strong acid or alkali solution, which the student should prove by experiment.

Our common starchy foods are prepared for use by **cooking or baking** in some manner. Application of heat produces several **interesting changes** in starch, some of which will be described below. One important effect is to make it digestible by saliva, which is practically without action on raw starch.

Ex. 13. Stir a small amount of uncooked starch into 5 Cc. of saliva, and allow to stand fifteen minutes at 35°–40°, and filter. Now apply the Fehling test, and note the absence of precipitated copper suboxide.

Action of Malt Extract on Starch.

In the germination of various grains, the enzyme known as diastase is formed, and this is exceedingly active in the conversion of starch. The reaction is largely employed in the arts, especially in the production of the sugar used by the brewer and distiller, and is brought about practically by the use of malt, which is commonly made from sprouted barley.

The barley is steeped in lukewarm water, and then spread out exposed to the air in a temperate atmosphere until it sprouts. The rootlet fibrils grow very rapidly, and after several days, when they have reached a certain length, the process of growth is checked by thorough drying in hot air kilns. The grain in this condition is rich in the starch converting element, and can be kept for use at any future time. When ground to a coarse meal, and mixed with water, the ferment present goes into solution along with other soluble products formed from the grain. When the solution is evaporated at a low temperature (best *in vacuo*) it yields the commercial product known as malt extract, which, however, is usually so carelessly made as to possess but little of the diastatic power of the original liquid.

Like the ptyalin of saliva diastase is destroyed by heat, and evaporation at any temperature above 70° C., would result in rendering the product valueless.

The following experiments show properties of the true malt extract :

Ex. 14. Mix about 10 Gm. of pale ground malt with 25 Cc. of lukewarm water, and allow the mixture to stand a short time, with frequent stirring or shaking. Then filter and add the clear, bright filtrate to a thin starch paste made of 10 Gm. of starch, with 250 Cc. of water. The starch paste must be cool when the malt extract is added. Now, place the mixture on the water-bath and warm to 50°-60° C., and maintain at this temperature. Note that the liquid gradually becomes thin and transparent. From time to time remove a few drops by means of a small pipette, and test with iodine solution. At first a deep blue color appears but this grows weaker, giving place to violet, then to yellowish brown, and finally no color is obtained, indicating completion of the reaction. The starch paste is first converted into dextrin and finally into maltose. Evaporate the solution to a very small volume and observe the taste and appearance of the residue. Save it for a later experiment.

Strong acids or alkalies and high heat destroy the diastase as with ptyalin, and in its behavior with raw starch malt extract resembles saliva.

Action of Extract of Pancreas on Starch.

As intimated above the pancreatic fluid contains a ferment which converts starch into sugar, maltose at first and by prolonged action into dextrose.

Several methods have been proposed for obtaining an active extract, but all require that the minced gland should remain a number of days in contact with the menstruum. The extracting liquid may be water, alcohol (dilute) or glycerol, and the reaction either acid or alkaline. It has not been found practically possible to separate by any of these processes the amylolytic from the other ferments present, that is from those which digest proteids, which emulsify fats and which curdle milk. A good extract which shows the characteristic reactions can be made in this manner:

Ex. 15. Cut a hog's pancreas into small pieces and pass these through a small sausage mill. Take about 10 Gm. of the finely divided matter and cover it with 50 Cc. of dilute alcohol of about 25 per cent strength. Allow the mixture to stand a week, filter it and evaporate the alcohol at a low temperature. Take up the residue with 50 Cc. of water, and use this solution for tests below.

Ex. 16. Prepare a starch paste with 5 Gm. of starch to 100 Cc. of water. Mix 10 Cc. of this paste, after cooling, with 5 Cc. of the pancreatic extract, warm to a temperature of $35°-40°$ and notice that the paste soon becomes thin and nearly clear. After a time test for sugar. Repeat the experiment, using pancreatic extract which has been boiled before mixing with the starch. The sugar reaction now fails to appear, showing that high temperature destroys the activity of the enzyme, as in the case of saliva.

Ex. 17. Instead of extracting the pancreas with weak alcohol, glycerol may be used. Take 10 Gm. of minced substance, cover with absolute alcohol in a small dish and allow to stand over night. Then pour off the alcohol and press out the residue between folds of filter paper. This treatment removes water. Now put the material in a test-tube and add 10 Cc. of pure glycerol. Cork and allow to stand a week. At the end of the time pour off the liquid and use a little of it for tests on starch paste. Save the rest for tests on proteids.

A few drops should be sufficient to give the sugar reactions above described.

Action of Heat on Starch.

An important article of commerce is made by heating starch to a temperature of about 250° C., and is called British gum, or dextrin. The same substance is produced at a lower temperature, usually at 125° to 150° by previously moistening the starch with very dilute nitric acid. The acid employed for this purpose need not be stronger than 0.15 of one per cent. The product made in this manner is lighter colored than that made in the dry way.

The dry preparation is illustrated in the following experiment:

Ex. 18. Heat about 10 Gm. of starch in a porcelain dish on a sand-bath to a temperature short of the point where it begins to scorch. It is necessary to stir well all the time, and continue the heat ten minutes after the starch has become uniformly yellowish brown. Then allow the dish to cool, add water and boil thoroughly, which brings part of the product into solution. When sufficiently diluted this solution can be filtered. The filtrate is precipitated by alcohol.

Dextrin, or British gum, with a small amount of water, constitutes a valuable mucilage or paste which has many uses in the arts. The solution, treated with extract of malt, yields maltose. Dilute acids bring about the same change, leaving dextrose, however, as the final product.

The Sugars.

Closely allied to the starches in origin, structure and chemical properties are the sugars, which, for our purposes, may be divided into two principal groups, the *glucoses* and the *saccharoses*. Representatives of both groups occur widely distributed in nature, in the animal as well as in the vegetable kingdom ; and the members of the second group, the saccharoses, can readily be converted into glucoses by treatment with dilute acids. For our purpose, in the glucose group we need consider only *dextrose*.

This occurs as grape sugar in many fruits, as diabetic sugar in the urine of persons afflicted with diabetes mellitus, normally in several animal tissues or secretions, in honey and elsewhere. As intimated above, it is formed as the final product in the complete digestion of starch by the enzymes of the body, and on the commercial scale is produced in immense quantities by the action of dilute sulphuric acid on starch. Illustrations of these processes have been given. We have now to consider some of the important chemical properties of dextrose.

REACTIONS OF DEXTROSE.

For experiments with dextrose, solutions made by dissolving crystallized commercial grape sugar in water may be used.

An important property of dextrose is its power of decomposing alkaline solutions of many metallic salts, usually with precipitation of the metal or a low oxide.

Ex. 19. Add to a dilute solution of dextrose enough copper sulphate solution to impart a very faint greenish blue color and then a considerable excess of strong potassium hydroxide solution. This produces no precipitation, but increases the color. On warming the solution a yellowish precipitate forms which grows bright red by boiling. This is cuprous oxide, and the test is known as *Trommer's test*. It is frequently employed to detect the presence of sugar in liquids, especially in urine, but on the whole is not as satisfactory as the next one or *Fehling's test*.

Ex. 20. The use of Fehling's solution has already been referred to. Its preparation is given in the appendix.

Add to a small volume of the sugar solution some Fehling solution, and as long as both liquids are cold and dilute no precipitate forms. On standing, however, in the cold, a precipitate will in time form. Precipitation takes place immediately on heating, the final reaction being the same as in the Trommer test.

The Fehling solution is in many respects the most valuable sugar test we have and is used quantitatively as will be described below.

In the method by the Trommer test if too much copper sulphate is added the color that appears on heating is black instead of red, from the formation of precipitated cupric oxide in place of cuprous, by reduction.

Ex. 21. Add to a dextrose solution some strong potassium hydroxide solution and then a very small amount of bismuth subnitrate. For an ordinary test a few milligrams will be enough. On boiling, a black precipitate appears which frequently forms a bright mirror on the walls of the test-tube. This precipitate seems to be a mixture of metallic bismuth with some oxide, and shows the strong reducing power of the sugar.

Ex. 22. Prepare an alkaline solution of mercuric cyanide as described in the appendix. Heat two or three cubic centimeters to boiling in a test-tube and add a small amount of sugar solution. A reduction of the reagent takes place with deposition of metallic mercury. This solution is sometimes employed in the quantitative determination of sugar, especially in urine.

Among the sugar reactions not depending on the reduction of metallic salts, three may be given as of especial interest. The first of these is the so-called fermentation test and may be briefly explained as follows:

Under the influence of yeast a number of saccharine substances have the property of splitting up into alcohol and carbon dioxide, when in sufficiently dilute condition. The practical and theoretical importance of this reaction is so great that it has been made the subject of a large number of investigations, which have established numerous points regarding the kinds of sugars which may be fermented, the by-products as well as principal products formed, the best conditions as to temperature and dilution of the fermenting solution, the isolation and peculiarities of various kinds of yeast and so on, but for our purpose a few experiments will show such points as may be looked upon as most worthy of attention.

The yeast ferment differs from the enzymes referred to above in being a living plant cell, capable of reproduction under proper conditions, but like the enzyme is active only within certain limits of temperature and strength of solution. Strong acids or alkalies destroy both.

Ex. 23. Rub up a quarter of a cake of compressed yeast with a little water to a thick cream, and then add about 100 Cc. of water. Allow the mixture to stand in a warm place some hours when it is ready for use. Now prepare a sugar solution by diluting 25 to 30 Cc. of commercial glucose syrup with an equal volume of water. Pour this solution into a flask holding about 100 Cc. and add about 25 Cc. of the yeast mixture. After mixing thoroughly determine the specific gravity. Close the flask with a good stopper through which a twice bent delivery tube passes, and set aside in a moderately warm place, the temperature of which should not go below 20° C. At the end of 24 hours note the escape of gas bubbles from the surface of the liquid. Now dip the delivery tube in lime water contained in a test-tube or small flask and observe the formation of a precipitate as the gas bubbles pass into the liquid. The precipitate is calcium carbonate. Allow the liquid to stand two days longer, that is about three days in all and observe that it has become much lighter in specific gravity and that it has the odor of alcohol. It is, in fact, a weak solution of alcohol, which may be obtained by distillation.

Ex. 24. Phenyl hydrazine test. A characteristic reaction of great practical value is given on the addition of phenyl hydrazine to a solution of dextrose under certain definite conditions. To a dilute dextrose solution add about a gram of phenyl hydrazine hydro-chloride, and two grams of sodium acetate. Heat on a water bath half an hour, and then allow the liquid to cool. There will now be found a beautiful yellow crystalline precipitate of *phenyl glucosazone*, the nature of which is best seen under the microscope. This test is one of great delicacy, and has been applied to the detection of traces of sugar in urine.

Another test which serves for the recognition of even minute traces of dextrose and other sugars, is the following, proposed by *Molisch :*

Ex. 25. To a small amount of a dilute sugar solution add two drops of a solution of α-naphthol, containing about 20 Gm. in 100 Cc. On shaking, the liquid becomes turbid. Now add to it an equal, or slightly greater volume of pure strong sulphuric acid and shake. A deep violet color appears, which gives place to a violet precipitate on addition of water. This reaction has been shown to be due to the combination of the α-naphthol with furfurol produced by the action of sulphuric acid on the sugar present.

SACCHAROSES.

Three sugars deserve our attention in the saccharose group, viz.: Cane sugar, milk sugar and malt sugar.

Cane sugar, or saccharose, occurs widely distributed in the vegetable kingdom, but on a large scale is obtained mainly from several varieties of cane, from the sugar beet, and from maple sap. It has not been produced synthetically.

Several reactions distinguish it from dextrose, as illustrated in the following experiment:

Ex. 26. Prepare a dilute solution of pure cane sugar and boil it with Fehling solution in the usual manner. Observe that no reduction of the copper compound takes place. Next boil a similar cane sugar solution with a few drops of dilute hydrochloric or sulphuric acid several minutes, neutralize with sodium carbonate, and then apply the Fehling test. The characteristic red precipitate now appears. In this reaction the cane sugar is broken up by the acid into a molecule of dextrose, and a molecule of lævulose, both reducing sugars.

Cane sugar solutions do not undergo fermentation directly with yeast, and differ from dextrose solutions also by their behavior with polarized light, which will be explained below.

Milk sugar resembles cane sugar in some of its chemical properties, but reduces copper solutions. It does not ferment with pure yeast. Dilute acids convert it into galactose and dextrose. A method of preparing milk for sugar tests will be given below. Tests applied directly are unreliable because of the presence of albuminoids.

The formation of malt sugar has been referred to above. It has few uses in the pure state, but is interesting as being one of the final products of the digestion of starch. Its action on Fehling's solution resembles that of dextrose.

The Determination of Sugars.

Numerous methods have been employed for the determination of sugars in solutions. Two will be described in detail here which depend on different principles and which are generally followed in practice. One of these methods is but a modification of the qualitative test carried out with the Fehling solution, while the other depends on the behavior of sugars with polarized light.

Method with Fehling's Solution.

Fehling's solution, as described in the appendix, is made arbitrarily of such a strength that one cubic centimeter is reduced by 5 milligrams of dextrose, on the supposition that the sugar and copper salt react on each other in the proportion of

$$5 \,(CuSO_4 \cdot 5\,H_2O) \text{ to } C_6H_{12}O_6.$$

It was formerly held that the reaction was a perfectly definite and simple one, and could be expressed in this manner, but it is now known that the dilution of the solutions is a very important factor in determining the amount of copper reduced. The best conditions to be employed in practice have been determined by *Soxhlet*, who found the reducing power of several sugars to vary as follows, when they were tested in solutions of one per cent strength:

0.5 Gm. of invert sugar in 1 per cent solution reduces 101.2 Cc. of Fehling solution, undiluted.

0.5 Gm. of invert sugar in 1 per cent solution reduces 97.0 Cc. of Fehling solution, diluted with 4 volumes of water.

0.5 Gm. of dextrose in 1 per cent solution reduces 105.2 Cc. of Fehling solution, undiluted.

0.5 Gm. of dextrose in 1 per cent solution reduces 101.1 Cc. of Fehling solution, diluted with 4 volumes of water.

0.5 Gm. of milk sugar in 1 per cent solution reduces 74 Cc. of Fehling solution, undiluted. The reducing power in diluted solutions is the same.

0.5 Gm. of maltose in 1 per cent solution reduces 64.2 Cc. of Fehling solution, undiluted.

0.5 Gm. of maltose in 1 per cent solution reduces 67.5 Cc. of Fehling solution diluted with 4 volumes of water.

The oxidizing power of one cubic centimeter of Fehling solution with each kind of sugar may be tabulated as follows, assuming the sugars to be in solutions of approximately one per cent. strength when acted upon.

One Cc. of Fehling solution oxidizes:

	When undiluted.	When diluted with 4 vols. of water.
Dextrose	4.75 Mg.	4.94 Mg.
Invert sugar	4.94 "	5.15 "
Milk sugar	6.76 "	6.76 "
Maltose	7.78 "	7.40 "

The practical application of the test is best shown by an experiment.

Ex. 27. Measure out accurately into a flask holding about 250 Cc., 25 Cc. of the copper solution and the same volume of the alkaline tartrate. Heat the mixture, or Fehling solution, on a wire gauze and note that it remains clear. Fill a 50 Cc. burette with a dilute dextrose solution and run 10 Cc. into the hot liquid. Boil two minutes, shaking the flask continuously, and allow the mixture to settle. If the supernatant liquid appears yellow it indicates that the sugar solution is much too strong and must be diluted with at least an equal volume of water before beginning another test. If, on the other hand, the liquid is still blue, add 2 Cc. more of the sugar solution, boil again for two minutes and allow to settle. If the color is now yellow an approximate value for the amount of sugar in the solution becomes known, but if still blue, the operation of adding solution and boiling must be continued until, after settling, a yellow color appears. Approximately 250 Mg. of dextrose is required to reduce the Fehling solution taken, and this must be contained in the sugar solution added. From this preliminary experiment calculate the amount of sugar present in each cubic centimeter.

Ex. 28. With the data obtained in the above experiment as a basis, make now a new sugar solution, having a strength of about one per cent. Measure out 50 Cc. of the Fehling solution, heat to boiling and run the new sugar solution from the burette as before, the first addition being about 20 Cc. Boil and note the color after settling and then cautiously continue the addition of sugar solution, a few tenths of a Cc. at a time, boiling after each addition, until the blue color gives place to a yellowish green and then, by the addition of a drop or two, to a pale yellow.

Under the conditions of this experiment 4.75×50 indicates the number of milligrams of sugar in the volume discharged from the burette. By dividing the product, 237.5, by this volume the number of milligrams of sugar per cubic centimeter is given.

If the sugar solution taken for analysis is very dilute the amount oxidized by each Cc. of the Fehling solution approaches five milligrams, as indicated by the table given above.

The method of employing the test as just outlined is sufficiently exact for the ordinary practical uses.

Sometimes the final disappearance of the copper from the solution is determined by filtering a few drops through a very small filter and adding a drop of acetic acid and a drop of ferrocyanide solution to the filtrate, when the characteristic color is given if a trace of copper is present.

With clear aqueous solutions this procedure can hardly be considered necessary as the operator can determine the end of the reaction with very great accuracy after a little practice. But in highly colored solutions or with diabetic urine the cuprous oxide precipitate settles very slowly at times, leaving the experimenter in doubt. It is then necessary to resort to the ferrocyanide or other test to discover the progress of the reduction. Under the head of urine analysis some modifications of the test will be given.

The direction is given above to shake the flask in which the mixture of Fehling solution and sugar is boiling. This precaution is necessary to prevent bumping. The neck of the flask is held by a test-tube holder (a strip of stiff paper is a good substitute) and a continuous rotary motion given to the liquid while it is being heated.

To determine cane sugar by the Fehling solution it must first be *inverted* or converted into a mixture of dextrose and lævulose. If the sugar is in the dry condition the inversion can be accomplished as follows: Weigh out 9.5 Gm., dissolve in 700 Cc. of water, add 20 Cc. of normal hydrochloric acid and heat for 30 minutes on the water bath. Then neutralize with 20 Cc. of normal sodium hydroxide solution and make up to 1000 Cc. on cooling. This gives now a one per cent solution which is employed as given for dextrose using the factor 4.94 instead of 4.75 as the amount of sugar oxidized by each Cc. of the copper solution. On the completion of the experiment take 95 parts of cane sugar for each 100 parts of invert sugar found.

If the cane sugar is in solution it must be diluted to a specific gravity of about 1.005 at 15° C., and then heated with acid, using the proportions given above.

Malt sugar and milk sugar are determined directly by the Fehling solution, using the proper factors as given in the table.

If the analyst has under examination a solution containing both cane sugar and dextrose the latter can be determined first as outlined above, and then the sum of the two sugars after the inversion of the saccharose. By subtracting the amount of Fehling solution required for a given volume of the original from that required for the same volume of the solution inverted, the equivalent of the cane sugar in the original is found, expressed as invert sugar.

Polarization Method.

Many substances, solids, liquids and gases, have the power of rotating the plane of polarized light passed through them, the amount, or number of degrees of rotation, being proportional to the number of molecules the light passes.

The observation of this phenomenon in solids and gases is usually a matter of some difficulty, and not, therefore, practically useful. But in liquids, single substances or solutions, the observation is one readily made, and has become the basis of an important method in the qualitative and quantitative examination of organic substances.

Instruments employed to determine the action of substances on polarized light, are, in general, called *polariscopes*, or *polarimeters*, and are made in a great variety of forms. For the details of construction of these instruments the student is advised to consult a special manual or work on general physics. Enough will be given here to render plain the construction and use of one instrument which can be used for practically all tests.*

In an ordinary ray of light the vibrations of the particles of ether producing it take place in all directions and perpendicular to the line of propagation of the light. The projections of these oscillations can be represented as shown in Fig. 5.

---

*Landolt's work, "The Optical Rotation of Organic Substances," is the standard authority on the subject.

Certain substances have the power of so changing a ray of light passed through, or reflected from them, that the oscillations no longer take place in all directions, but in certain definite directions only. Fig. 6 represents these vibrations as taking place in two directions, in a vertical

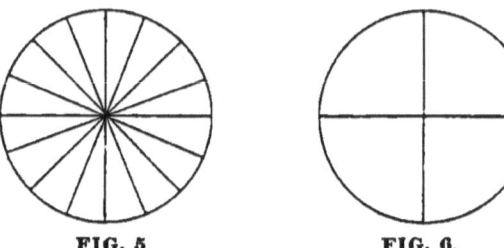

FIG. 5.   FIG. 6.

plane, and in one at right angles to it. Light so modified is spoken of as "plane polarized light," and for most purposes is produced by the passage of an ordinary beam of

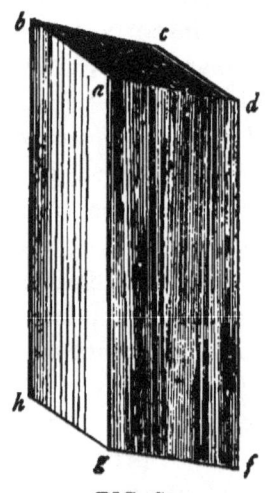

FIG. 7.

light through a specially constructed prism of Iceland spar known as a Nicol's prism. The construction and behavior of this prism can be shown by the accompanying diagrams.

Fig. 7 represents the natural crystal of Iceland spar and a line drawn from d to h represents the principal axis. A plane through $b\ d\ h\ f$ is a principal plane with the angles $d\ b\ h$ and $h\ f\ d$, Fig. 8, $= 71°$ nearly. If we suppose a ray of light to fall on the surface $a\ b\ c\ d$ it will pass

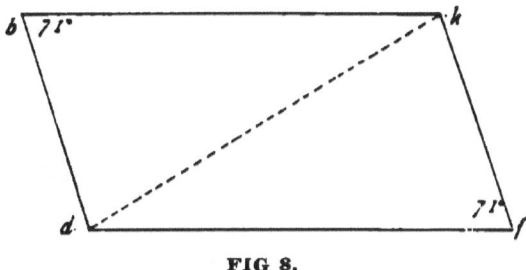

FIG 8.

through and emerge as two rays, an *ordinary* and an *extraordinary* as shown in Fig. 9.

The light here has suffered double refraction and the ray which is bent the most from its course, $o\ p$, is the ordi-

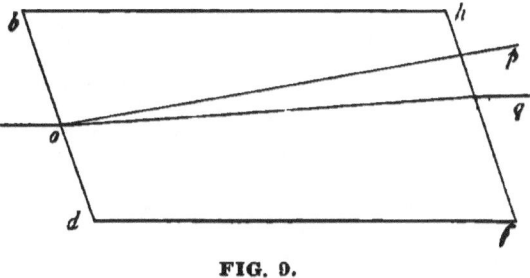

FIG. 9.

nary ray and the other, $o\ q$, the extraordinary. These rays are found by experiment to be both polarized, one in a plane parallel to the principal plane or section of the crystal $b\ d f h$, and the other at right angles to it. It is assumed that the ordinary ray is the one polarized in the direction parallel to the principal section.

For the purpose before us, it is necessary to eliminate one of these rays, best the ordinary, and this is accomplished in the following manner. The four faces *a b c d* and *e f g h* of the crystal are ground down until the angles *h b d* and *h f d* of the principal section are not 71° but 68°, which leaves the angle *b d h*, as shown in Fig. 10, 90°. Now

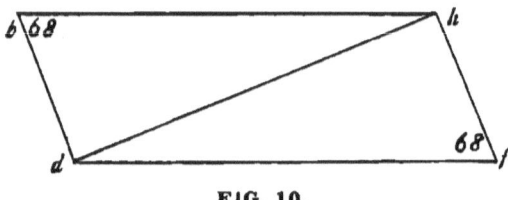

FIG. 10.

the crystal is sawed in two halves, the cut passing through *d* and *h* and perpendicular to the plane *b d h f*. The cut surfaces are covered with Canada balsam and pressed together again, which gives us a crystal exactly like the original one in appearance, but with somewhat different physical properties.

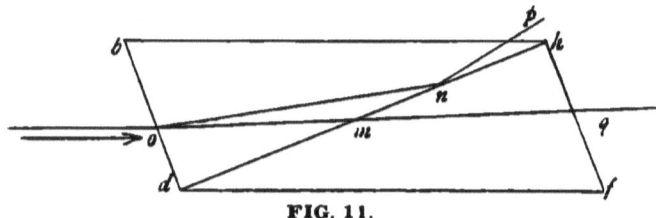

FIG. 11.

The light enters the "Nicol" at *o*, Fig. 11, suffers double refraction as before, giving the rays *o m* and *o n*. The extraordinary ray, *o m*, passes through the balsam surface *d h*, and emerges in the direction *q*, while the ordinary ray fails to pass the balsam, but is thrown to one side as it strikes it at a very oblique angle, and emerges from the prism in the direction *p*. The ray at *q* is polarized but its intensity is, of course, only one-half of that of the original ray at *o*.

Suppose, now, that two of these prisms are placed in positions as shown in the following figure, Fig. 12, with their principal sections in the same plane, that is perpendicular to the plane of the paper.

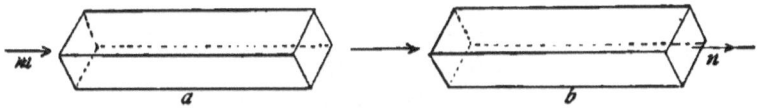

FIG. 12.

A ray of light entering *a* is polarized and emerges as a ray of half the intensity, the plane of polarization being parallel to the plane of the paper. This ray now passes into *b*, the position of whose planes is symmetrical with those of *a*. The light then emerges at *n* practically as it entered *b*, that is, as an extraordinary plane polarized ray and of nearly half the intensity at *m*.

But, on the other hand, if the prism *b* is turned so that its principal section is in a plane perpendicular to that of *a* we have the condition shown in Fig. 13.

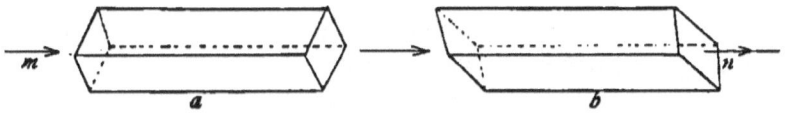

FIG. 13.

The extraordinary plane polarized ray leaving *a* enters *b* and behaves here as an ordinary ray until it reaches the balsam layer. Here it suffers total reflection and is thrown to one side, no light whatever emerging in the direction *n*. If the prism *b* has an intermediate position with reference to *a* any ray entering it from *a* is broken up into two rays, an ordinary and an extraordinary, of which the former is always lost by reflection from the balsam layer. If the prism *b* is placed so that its principal plane makes an angle of 45° with that of *a* the ordinary and extraordinary rays have equal intensity and the latter emerging finally in the direction *n* has just half the intensity of the light entering *b*, or one-fourth the intensity of that entering *a*. From this it follows that by giving *a* a fixed position and moving *b* around a horizontal axis parallel to the plane of the paper the light leaving it in the direction *n* passes through all degrees of intensity between that of the extraordinary ray of *a* and zero. It would further follow that with the two prisms placed as in Fig. 13, that is so that no light passes both, their principal

FIG. 14.

sections being at right angles to each other, a ray of some intensity can still be made to emerge from *b* after leaving *a*, provided we place between the two something which is capable of rotating the extraordinary polarized ray before it reaches *b*. The rotation of the plane of vibration of this ray to the right or left would be equivalent to the rotation of *b* to the left or right through the same number of degrees.

Now, there are many substances, solids, liquids and gases, which have the power of twisting or rotating the polarized ray in this manner, and the application of polarizing apparatus to the study of chemical phenomena depends on this fact.

For convenience in handling, the Nicol prisms are fastened in tubes of brass by the aid of pieces of blackened cork. The mounted prism shows in section as Fig. 14.

Suppose now we support two of these Nicols on a

stand, and place between them a glass vessel with plane parallel glass walls.

Allow a ray of light to pass in the direction indicated. If the glass vessel is filled with clear water, and the Nicol *b* is placed in position symmetrical with *a*, the light leaving *a* will pass through the water and through *b*, as above in Fig. 12. If *b* is placed as in Fig. 13 no light will pass through; the water is, therefore, without action. But if instead of with water we fill the glass vessel with a sugar solution, or with a solution of Rochelle salt, we observe now that some light does pass through *b*, and that to reach the condition of darkness again it will be necessary to rotate the latter prism through a certain number of degrees.

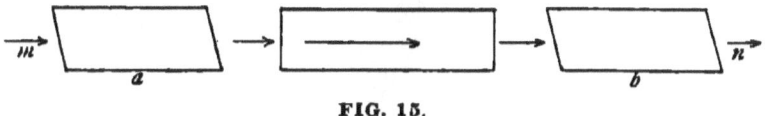

FIG. 15.

By trying different solutions in the glass vessel it will be further noticed that the number of degrees through which *b* must be rotated to secure perfect darkness, varies not only with the nature of the solution, but with its concentration.

In this combination of prisms *a* is called the polarizer, and *b* the analyzer, because it permits us to detect and measure the phenomena of "rotary" polarization.

The fact that a number of organic liquids and other organic bodies in solution have the power of rotating the plane of polarized light was discovered by the French physicist, *Biot*, and at his suggestion several forms of apparatus were constructed with which this phenomenon could be observed. These pieces of apparatus are in general called polariscopes, or polarimeters, and in their simplest form consist essentially of a polarizing and an analyzing prism, a tube placed between them to hold the liquid under investigation and some arrangement to measure the number of degrees through which the analyzer may be rotated.

Fig. 16 shows a stand arranged to hold two crystals or two Nicol prisms, with which the general phenomena of

FIG. 16.

polarization may be studied, the rotation of one of the prisms, as B, being measured on the graduated circle, C, while Fig. 17 shows one of the first forms employed in the practical examination of liquids.

The original Mitscherlich apparatus, shown in Fig. 17, contains nothing more than this: a stout brass pillar supports a horizontal arm d, on one end of which is a brass tube containing the polarizing prism, *a*. At the other end of the arm is a graduated circle fixed in position. In the center of this is an opening in which rotates the tube holding the analyzing prism, *b*. To this tube a handle, *c*, is attached, and also a pointer moving over the graduated circle. Between the two prisms a glass or metal tube, *f*, with ends of plane glass plates perfectly parallel to each

FIG. 17.

other is placed. This tube may be 100 to 600 Mm. in length, and holds the liquid to be examined.

In making an observation the apparatus is directed toward a bright light, and the analyzing Nicol turned until the condition of greatest darkness is reached. The tube, *f*, with the plate at one end in place, is now filled with the liquid to be tested, and the other glass end-plate screwed down by means of the brass cap in such a manner as to exclude all air bubbles, leaving the tube quite full and clear. It is placed in position, and if it contains a rotating substance it will be found necessary to turn the analyzer through a certain angle, as indicated by the movement of the pointer over the graduated circle, to obtain the darkened field again. The number of degrees read off is called the angle of rotation, and is characteristic of the substance, but depends on the concentration of its solution and the length of the tube.

This form of instrument cannot be used for exact quantitative measurements, but only for qualitative indications. More complete instruments have been devised by *Wild*, *Laurent*, *Jellett*, *Landolt*, and others, some of which can

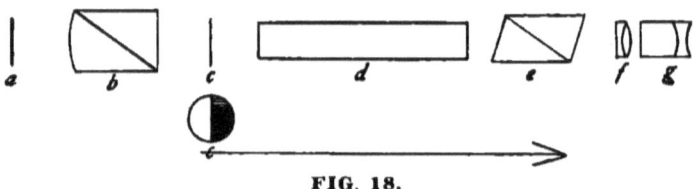

FIG. 18.

be used for polariscopic observations in general while others are employed only in the examinations of saccharine liquids. A very useful form suitable for all kinds of observations with homogeneous light is the *Laurent*, as made by several German and French firms.

The form made by *Schmidt & Haensch*, of Berlin, is shown in Fig. 19, while the arrangement of the optical parts is illustrated in Fig. 18.

Fig. 20 shows the lamp by which the yellow light is produced, sodium carbonate or bromide being volatilized in the flame.

Homogeneous yellow light is supposed to pass from *a* to *g*, Fig. 18. It first meets, at *a*, a thin plate of potassium bichromate, held between two glass plates. The ob-

ELEMENTARY CHEMICAL PHYSIOLOGY. 49

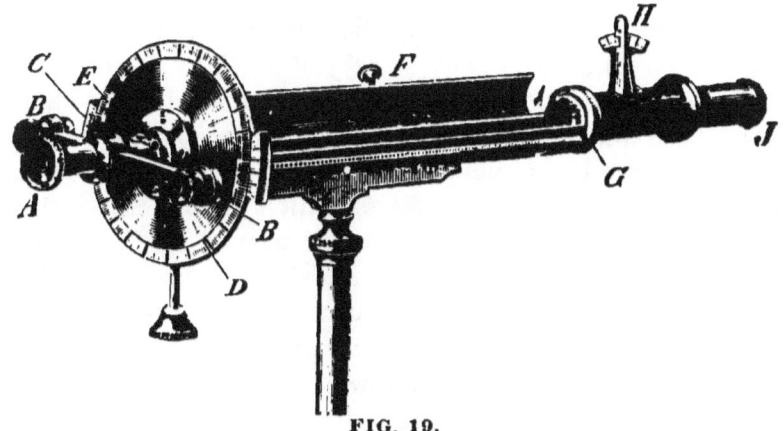

FIG. 19.

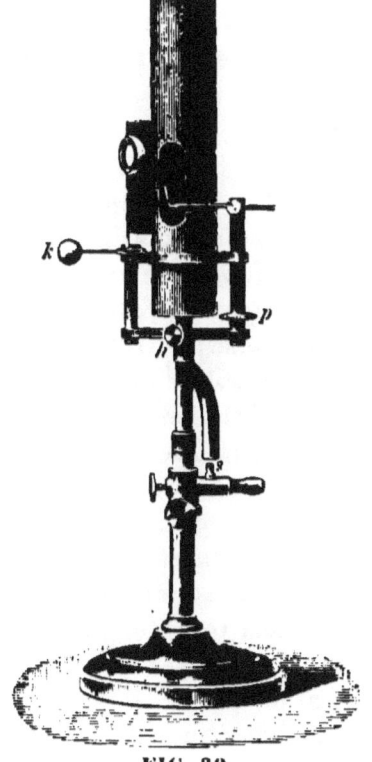

FIG. 20.

ject of this is to further purify the sodium light by absorbing any foreign rays. The pure yellow light then passes into the polarizing Nicol b, which is mounted in a short brass tube in such a manner that it can be slightly rotated by means of a lever shown at H, of Fig. 19. At c there is a round glass plate over half of which is mounted a thin plate of quartz shown by the shaded part of c. This plate of quartz is ground parallel to the axis of the crystal and is of such a thickness that it produces a difference of half a wave length in the rate of the two polarized rays of yellow light passing through it. Following it is the tube d to hold the liquid to be examined, then the analyzing Nicol e, and finally the achromatic telescope eyepiece made up of f and g. The analyzing Nicol can be rotated around its center and is firmly mounted in the middle of a graduated circle which turns with it.

The part of this instrument which distinguishes it essentially from the simple Mitscherlich and other forms is the small quartz plate at c. When the polarizing prism b is adjusted so that the plane of polarization is exactly parallel to the axis of the quartz plate on c (or its vertical edge) the two halves of c appear equally light or dark when viewed through the analyzer; with the latter at right angles to the polarizer there is absolute darkness. But if the adjusting lever over b is turned slightly so as to rotate b, and therefore the plane of polarization, with reference to the vertical line of the quartz plate on c, then there is no condition of absolute darkness for any position of the analyzer, but a condition approaching darkness with the shadow the same on both halves of c when the analyzer is turned to cross the polarizer. The slightest rotation of the analyzer to the one side or the other now causes one-half of c to grow suddenly dark and the other light. Reversing the direction of the rotation of the analyzer causes a reversion of light and shade on viewing c.

It is therefore very easy to determine the position of the analyzer for greatest darkness, something which was not possible with the old Mitscherlich instrument. The addition of the small quartz plate is merely a device to enable the observer to find accurately and quickly the zero point of the instrument. In other modern instruments the same end is reached by other means, although not always so perfectly as here.

The amount of rotation for any substance depends on the length of the column taken, on the concentration and density of the liquid and to a slight extent on temperature. Each substance is characterized by a *specific rotation* which may be defined as the rotation of a liquid column one decimeter in length, each cubic centimeter of which is assumed to contain one gram of active ingredient.

The following designations are commonly employed:

a = observed angle of rotation.

l = length of tube in millimeters.

c = concentration of liquid, that is number of grams in 100 cubic centimeters.

d = density of liquid polarized.

p = percentage strength of liquid, that is the number of grams of active substance in 100 grams of liquid or solution.

$c = p \cdot d.$

$[a]$ = the specific rotation.

The specific rotation of an active liquid—oil of turpentine, for instance—is expressed as follows:

$$[a] = \frac{100\, a}{l.\, d.}$$

If the liquid under examination is a solution of an active substance the above formula cannot be used, but the following, as the percentage strength must be taken into consideration:

$$[a] = \frac{10^4\, a}{l.\, p.\, d.}$$

In using this formula it is necessary to know both the percentage strength and the specific gravity, but as $p.\,d$ is the concentration of the solution, that is the number of grams in 100 Cc., a result equally valuable for many purposes is reached by weighing out the active substance in grams and dissolving it to make 100 Cc. of solution.

In this case the formula to be used is

$$[a] = \frac{10^4\,a}{l.\,c.}$$

As an illustration of the use of the polariscope determine the specific rotation of several sugar solutions.

Ex. 29. Weigh out 10 Gm. of cane sugar and dissolve it in distilled water in a 100 Cc. flask. When solution is complete, fill to the mark and shake well.

Close one end of the 200 Mm. polarizing tube with its glass cap and fill with the solution to be tested. Now put on the other glass end so as to exclude all bubbles of air and screw it down. Place the filled tube in its receptacle in the polariscope and direct this toward a bright sodium light in an otherwise dark room.

If the instrument had been adjusted so that the zero of the scale corresponded to the condition of equal shadows on the circular plate it will be found that with the sugar solution between the polarizer and analyzer, the latter, with the divided circle, must be rotated through a certain number of degrees to reach the same shadow effect. Note the direction and amount of this rotation, and calculate the specific rotation according to the last formula above.

Ex. 30. Repeat the experiment, using solutions of grape sugar and honey in known amount. The rotation of the latter varies with different samples.

The following are mean values of the specific rotations of different substances.

D, in these formulas, indicates sodium light.

Saccharose................ $[a]_D =$ 66.5°, $c = 10$—30
Lactose ($C_{12}H_{22}O_{11}.H_2O$)... " $=$ 52.5°, $c = 5$—12
Dextrose (anhydrous)...... " $=$ 52.7°, $c = 2$—20
Invert sugar.............. " $=-27.9°$, $c = 17$—
Serum albumin............ " $=-56.°$
Egg albumin.............. " $=-35.5°$

The use of the polariscope in the special examination of proteids will be referred to later. The instrument is employed also in many chemical investigations in lines not related to physiological chemistry.

In illustration of the methods of calculation the details of some observations will be given. A sample of pure oil of turpentine was polarized in a tube 300 Mm. long and at the temperature of 20° C. The rotation of the plane of polarization was found to be 40.504°. The specific gravity of the oil, accurately determined, was found to be at 20°, 0.8650, water at 4° being taken as unity. We have therefore,

$$a_D = 40.504$$
$$l = 300.$$
$$d = 0.865$$

from which follows

$$[a]_D = \frac{100 \times 40.504}{300 \times 0.865} = 15.608°$$

In another case 25 Gm. of pure Rochelle salt was dissolved in distilled water and made up to a volume of exactly 100 Cc. at 20°. The solution was examined in a tube 200 Mm. long and was found to have a rotation of 11.058°.

Here we have

$$a_D = 11.058°$$
$$l = 200.$$
$$c = 25.$$

from which follows

$$[a]_D = \frac{10^4 \times 11.058}{200 \times 25} = 22.116°$$

Fermentation of Sugars.

It has been shown above that certain sugars in solution readily undergo fermentation by action of the yeast cell. The products formed are essentially ethyl alcohol and carbon dioxide.

Cane sugar does not ferment with ordinary yeast, and strong solutions cannot be fermented at all. In fact, many of the uses of cane sugar depend on its power of preventing fermentations or putrefactions. This is illustrated in the "canning" or preservation of fruits, and in the employment of syrups in the pharmacopœia.

Ex. 31. Make a strong cane sugar syrup, by boiling or heating together 10 Gm. of sugar and 10 Cc. of water. Allow to cool and add about a gram of crumbled compressed yeast, and then set aside for several days. The solution should be found free from any signs of fermentation.

Conditions of Alcoholic Fermentation.

The following experiments will be of interest in showing some of the conditions on which alcoholic fermentation depends.

Ex. 32. Prepare a twenty per cent solution of commercial dextrose and pour 20 Cc. of it in a test-tube. Add some yeast and allow to stand about two days in a moderately warm place. At the end of this time it should be found in active fermentation, as shown by the escape of gas bubbles and the odor of alcohol.

Keep for future experiment.

Ex. 33. Prepare a tube with sugar solution and yeast as in the last experiment. Close it loosely with a plug of absorbent cotton and heat to boiling, allowing steam to escape through the cotton. If the tube is now left to itself for several days it will be found that fermentation has not taken place, showing that heat destroys the characteristic property of the yeast cell.

Ex. 34. Prepare another tube with sugar and yeast and add 10 Cc. of strong alcohol. Shake the mixture and allow to stand. No fermentation appears as the activity of the yeast cell is destroyed by alcohol. We have good familiar illustrations of this in the self-preservation of certain "heavy" wines, as ports, sherries and malagas, while "light" wines, which contain ten to twelve per cent of alcohol, usually must be kept tightly bottled for preservation.

Ex. 35. Tests for alcohol. We have many tests by which the presence or formation of alcohol may be shown. The fermentation of a saccharine liquid is followed by a lowering of the specific gravity as already mentioned, and a practical quantitative test is based on this fact. A simple chemical test for the presence of alcohol, which in most cases is sufficient, is the following: Add to the clear liquid to be examined a few small crystals of iodine, warm to about 60° C., and then add enough sodium hydroxide or carbonate to produce a colorless solution. An excess of the alkali must not be used. In a short time bright yellow crystals of iodoform precipitate, easily recognized by their color and odor. Certain other liquids give the same test.

Acetic Acid Fermentation. Weak alcoholic solutions, if containing small amounts of foreign matter, especially nitrogenous, become sour, if exposed to the air, from formation of acetic acid. This is due to the fact that a small microscopic plant cell, the *mycoderma aceti*, is generally present in the atmosphere, and in sufficient numbers to oxidize the alcohol to acid.

Ex. 36. The solution of Ex. 32 becomes sour if allowed to stand four or five days in a warm place. When the odor of alcohol has practically disappeared add some precipitated chalk and evaporate to dryness. Extract with a little water, filter and concentrate the filtrate to a small bulk; add equal volumes of alcohol and strong sulphuric acid and apply heat. The characteristic odor of acetic ether is developed, proving the presence of acetic acid as a product of fermentation.

The acetic acid ferment is destroyed by high temperature or strong alcohol, which can be shown by experiments similar to those detailed above.

Lactic Acid Fermentation. By the action of a peculiar ferment solutions of sugar in presence of nitrogenous matters become converted into lactic acid, according to the equation:

$$C_{12}H_{22}O_{11} + H_2O = 4\,C_3H_6O_3$$

The reaction takes place best in neutral or slightly alkaline solution and is speedily stopped by the accumulation of acid formed. In practice, therefore, it is customary to add some chalk or zinc oxide to the fermenting mass to take up the acid as fast as formed. The fact of the production of acid is shown by the following experiment:

Ex. 37. Dissolve 30 Gm. of cane sugar in 150 Cc. of boiling water and add about 30 Mg. of tartaric acid. After this mixture has stood two days, add some putrid cheese, about half a gram, and 40 Cc. of sour milk. Then add 15 Gm. of zinc oxide and allow the mixture to stand ten days in a warm place where the temperature is about 45° C., stirring frequently. At the end of this time heat to boiling, filter hot and allow the zinc salt to crystallize out on cooling. Dissolve the crystals in a small amount of hot water, and allow them to settle out again. Collect these purer crystals, mix with water and pass hydrogen sulphide into the mixture in excess. Filter off the precipitated zinc sulphide and shake the aqueous filtrate with ether in two portions, using 25 Cc. in all. Allow the ether to evaporate spontaneously, leaving the lactic acid as a thick liquid.

Butyric Acid is produced from lactic acid by prolonged fermentation.

If the fermented mass in the above experiment, containing crystals of calcium or zinc lactate, is allowed to stand some weeks at about the same temperature, bubbles of hydrogen and carbon dioxide will escape, and decomposition, according to the following equation takes place:

$$2 C_3 H_6 O_3 = C_4 H_8 O_2 + 2 C O_2 + 2 H_2$$

This action, like the former, is due to the presence of a minute microörganism.

Glycogen or Animal Starch.

Closely related to the carbohydrates described above we have a compound produced in the animal body and known as *glycogen*. This substance is found in small quantities in many of the tissues, but is relatively abundant in the liver only, from which organ it is usually prepared. It is also found in the common oyster, which is sometimes employed as a starting point in the preparation of the pure substance.

The laboratory production of glycogen is best carried out as follows:

Ex. 38. Kill a rat or a rabbit; remove the liver as quickly as possible and without delay cut it into small bits which throw into a vessel of boiling water. The weight of the water should be about five times that of the minced liver. Boil five minutes, then remove from the water and rub up in a mortar with fine clean quartz sand. In this way the fragments of liver become thoroughly disintegrated. The contents of the mortar, sand as well as liver, are thrown into the boiling water again and kept at 100° C. fifteen minutes. At the end of this time enough dilute acetic acid must be added to impart a faint acid reaction. This coagulates and precipitates some albuminous matters which are separated when the hot mixture is filtered.

In the opalescent filtrate, which must be collected in a cold beaker, a further precipitation of albuminous matter is effected by adding a few drops of hydrochloric acid and some potassium mercuric iodide as long as a precipitate forms.

Filter again and use the dilute aqueous solution of glycogen resulting for tests below.

Ex. 39. Evaporate about half of the liquid above to a small volume and precipitate impure glycogen as an amorphous white powder by addition of strong alcohol.

Ex. 40. Add a little tincture of iodine to a small portion of the solution, and note the red color produced. This color is discharged by heat. Boil some of the solution with dilute hydrochloric acid ten minutes; neutralize the acid nearly, cool and again test with iodine. No color is now produced, as glycogen has disappeared under the treatment, having been converted into sugar.

After death the store of glycogen in the liver rapidly disappears, so that tests applied at the end of a day or two fail to show its presence.

Ex. 41. Cut some common beef liver from the market into small bits and extract with boiling water. Boil longer to coagulate the albuminoids, after adding some sodium sulphate. Apply the iodine test for glycogen, which is found absent, and the Fehling test for sugar, which is found present in quantity.

In Experiment 38, the direction is given to treat immediately with boiling water. This is done to destroy a ferment-like body in the liver, which speedily converts glycogen into sugar.

Solutions of glycogen in water exert a marked dextrorotatory action on polarized light, but observers differ as to the exact specific rotation. It is about

$$[\alpha]_D = 210°.$$

Chapter III.

FATS.

THE bodies commonly called fats are compounds of glycerol with palmitic, stearic and oleic acids. Butter fat contains in addition glycerol compounds of butyric, caproic and other less important acids. The fats of the vegetable kingdom are in many respects similar to those of the animal kingdom, and like the latter, serve as food for man.

The important animal fats are butter, lard and the several kinds of tallow. The important vegetable fats are olive oil, cottonseed oil, sesame oil, peanut oil, cocoa butter and others. Pure fats are usually separated from the accompanying animal or vegetable tissue by pressure, or by melting, as illustrated by the manufacture of olive oil and the rendering of lard or tallow.

Origin of Fats in the Body. The fats found in the animal tissues are obtained from three general sources. First. They are taken directly from the fatty matters of the food, and the amount so absorbed may be considerable, as has been shown by several physiologists who have investigated the question.

Second. A production of fat goes on in the body when the animal is fed on a diet of albumin alone. The amount of butter which a cow will produce is far in excess of the fat in the feed, and can be accounted for only on the assumption of decomposition of albuminous matters. Dogs have been starved until their bodies were practically free from fat and then fed on muscle and soaps of palmitic and stearic acids. After a time they were killed, and on analysis their bodies were found to contain not only palmitin and stearin, but also *olein*. This must have come from a breaking down of the food proteids. The formation of large quantities of *adipocere* in cadavers is considered, also, as a proof of the production of fatty acids from albuminoids.

Finally, it has been abundantly proven that carbohydrates give rise directly to fats by molecular changes in the body. *Indirectly*, the consumption of carbohydrates favors production of fats by taking the place of the albuminoids in oxidation. The latter are, therefore, left to be broken up into a nitrogenous residue, and a fat-forming residue which is stored up in the tissues.

Some of the important properties of fats are shown by the following experiments :

Ex. 42. Note the solubility of small bits of tallow in ether, chloroform, benzine and alcohol, using in each case the same volume of liquid, with equal weights of fat. The solubility in alcohol will be found much less than in the other menstrua. The common fats are insoluble in water.

All fats undergo a very remarkable change when heated with solutions of caustic alkalies. This treatment is sufficient to split them up into their component parts, glycerol on the one hand and the alkali-metal salt of the fatty acid on the other. Thus stearin, or glyceryl stearate, the compound of glycerol with stearic acid, yields, when treated with a strong solution of potassium hydroxide, free glycerol and potassium stearate, as shown by the following tests:

Ex. 43. Boil about 25 Gm. of cottonseed oil or tallow with a solution of 10 Gm. of potassium hydroxide in 25 Cc.

of water. Stir the mixture thoroughly until it becomes homogeneous, that is, until no oil globules are seen floating on top of the aqueous liquid, which may require half an hour. Add water from time to time, to make up for that lost by evaporation.

The resulting mass is a mixture of glycerol, excess of alkali and soft soap.

Ex. 44. The presence of fatty acids in the above mixture can be shown by adding to a portion of it enough hydrochloric acid to decompose the soap. Use about half the product of Ex. 43, dilute it with water, and add the acid in slight excess, about 10 Cc. of the strong commercial acid. Warm on the water-bath, which will cause the liberated fatty acids to collect on the surface as a liquid layer as soon as the temperature becomes high enough. Add more water and allow the whole to cool. A semi-solid layer of fatty acids can now be lifted from the surface of the liquid. The hardness of the mixed acids depends on the nature of the fat taken for experiment. Mutton and beef tallows yield very solid acids; with lard the mass is softer, while with some oils the acids do not solidify at all at the ordinary temperature.

Ex. 45. Dissolve a small portion of the fatty acids in warm alcohol, nearly to saturation. On cooling, the acids separate in crystalline scales.

Ex. 46. The presence of glycerol as one of the products formed by the saponification of fats is best shown as follows: Mix 50 Cc. of cottonseed oil with 25 Gm. of litharge and 100 Cc. of water in a porcelain dish. Place over a Bunsen burner on gauze and stir until all oil globules have disappeared, adding a little water from time to time. The litharge with water acts as lead hydroxide and saponifies the fat, forming an insoluble lead soap, or plaster, and glycerol. When the saponification is complete add more water, heat and stir well to dissolve the glycerol, allow to settle a short time and pour the aqueous solution through a filter. To the residue add water again, heat, allow to settle and pour through the same filter. Concentrate the mixed filtrates to a small volume and after cooling observe the sweet taste of the thickish residue.

Glycerol of commerce was at one time made by this slow process, but at the present time it is made by other reactions from fats, usually by the action of superheated steam. This passed into melted fat effects a separation into glycerol and fatty acid. Sometimes the fat is decomposed by milk of lime and the glycerol separated by a current of superheated steam with which it is volatile.

The inverse process of combining glycerol and fatty acids to form fats is not so easy, but it has been done on an experimental scale, yielding not only the tri-acid compounds similar to those existing in nature, but mono- and di-acid compounds as well.

Fats being insoluble in water, a permanent mixture of the two cannot be obtained directly. A separation always follows after a time. By the action of certain substances, however, the fat can be brought into a very finely divided condition called an emulsion, which has important properties.

Emulsions.

Fats can be thrown into this peculiar physical modification by a variety of additions, which are illustrated in the experiments below. In order that a fatty substance may leave the alimentary canal and enter the circulation it is necessary that it must first be brought into a condition in which it can be absorbed by the lacteals. Emulsification accomplishes this, which is mainly carried out in the small intestine.

Ex. 47. Add to 5 Cc. of cottonseed oil half its volume of strong white of egg solution and shake thoroughly. The liquids mix and form a white mass or emulsion.

Ex. 48. To 5 Cc. of cottonseed oil containing a little free fatty acid add 10 drops of strong sodium carbonate solution and shake. A good stable emulsion is made in this way.

Ex. 49. In the intestine this reaction is brought about by the action of the pancreatic juice and bile, which can be illustrated in this manner: Pour about 10 Cc. of refined cottonseed oil in a warm mortar and add two or three grams of fresh finely divided pancreas. This can be obtained by running the pancreas through a good sausage mill. Rub the mixture thoroughly with a pestle, by which means a white milky mass is obtained.

To produce an emulsion the presence of free fatty acids seems to be necessary. Now it is probable that the pancreatic juice contains a product capable of splitting fats into free acid and glycerol, the first of which with the alkaline juice produces the emulsion. It is certain that the emulsifying power of the juice does not depend on the presence of an enzyme, because emulsification takes place with boiled pancreatic juice as well as with fresh. In its work of splitting fats the pancreatic juice is assisted by the bile, some tests on which will be given later.

Crystallization of Fats.

The neutral fats as well as the fatty acids readily assume the crystalline form under certain conditions. Fats in the tissues are undoubtedly free from crystalline structure, and when melted out at a low temperature or separated by pressure the amorphous form is the common one. Melted fats, on standing, generally become crystalline, which can be shown by melting a little tallow and placing a drop of it on a glass slide to cool. If a cover glass is pressed down on the fat before it becomes hard this will give a flat field in which the crystalline form can be seen under the microscope.

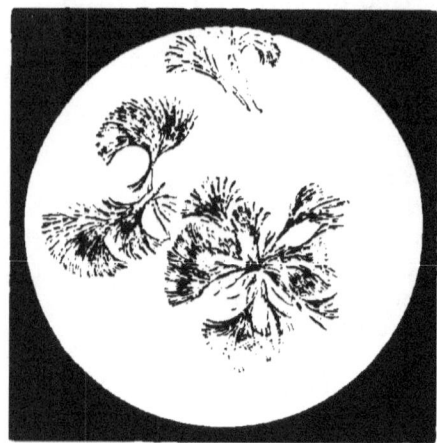

FIG. 21.

Human fat, crystallized from chloroform.

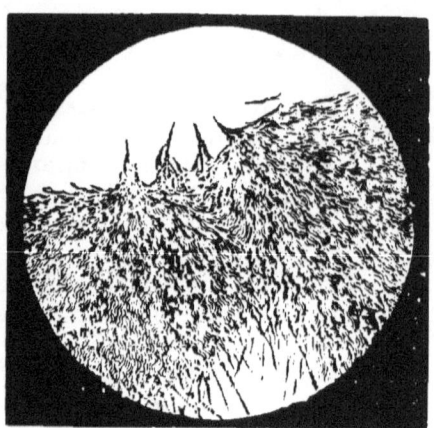

FIG. 22.

Dog fat, crystallized from chloroform.

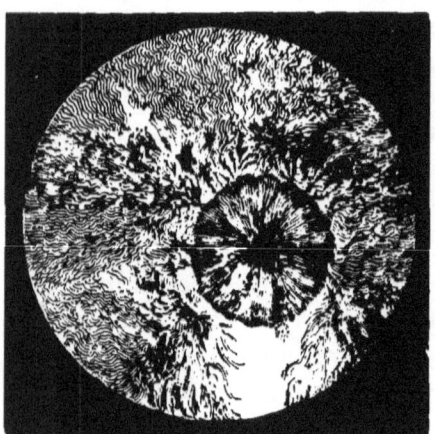

FIG. 23.

Cat fat, crystallized from chloroform.

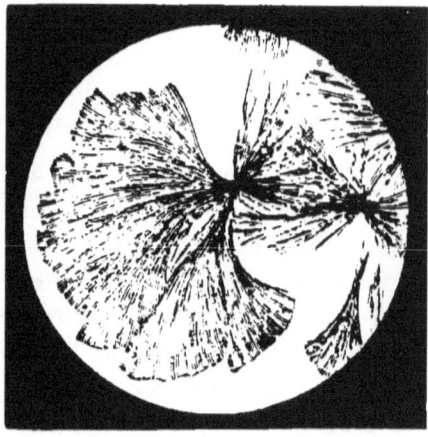

FIG. 24.

Beef tallow, crystallized from chloroform.

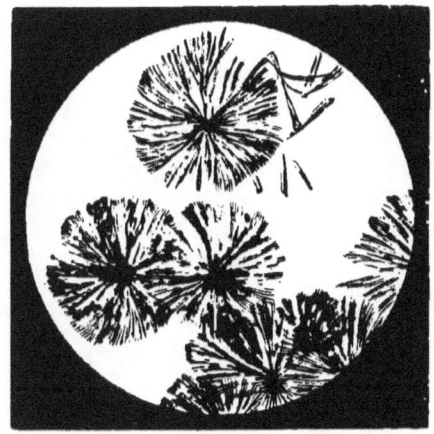

FIG. 25.

Beef tallow, crystallized from chloroform.

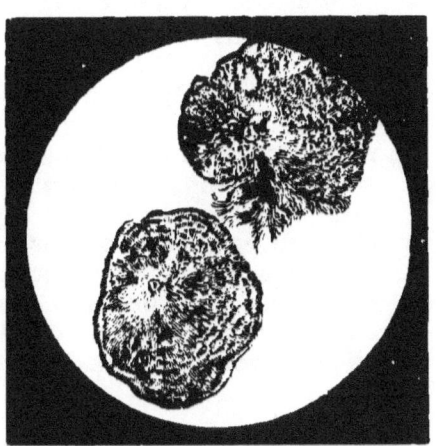

FIG. 26.

Pure stearin, crystallized from chloroform.

FIG. 27.

Mutton tallow, crystallized from chloroform.

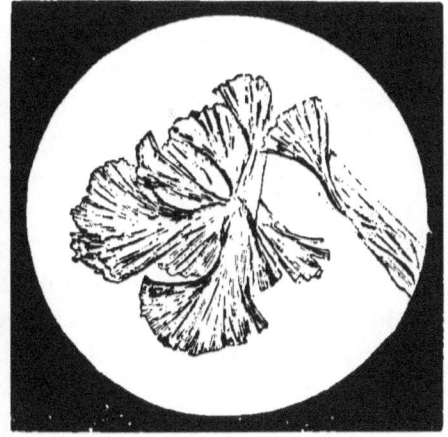

FIG. 28.

Lard, crystallized from chloroform.

Perfect crystallization can be secured by deposition from a menstruum.

Ex. 50. Dissolve some mutton or beef tallow in chloroform and with a glass rod put two or three drops of the nearly saturated solution on the center of a glass slide. As the chloroform evaporates a film begins to form on the top of the drop. Now put on a perfectly clean cover glass and allow to stand until crystallization is complete, which may require only a few minutes or some hours, the time necessary depending on the temperature and on the concentration of the solution. Examine the crystals with a microscope. Use a power of 200 to 300 diameters.

Very peculiar crystallizations are shown by certain fats as can be seen by the annexed cuts. Some of these forms have been taken as characteristic and have been considered as sufficient for the identification of several fats, even in mixtures.

It is true that in general the crystals of mutton tallow or beef tallow are different from those of pure lard, but it is also true that mixtures can be made which yield crystals practically identical with those usually seen in the fields from lard. At various times it has been supposed that adulterations of lard or butter could be demonstrated by these microscopic appearances, but, one by one, the points brought forward as characteristic have been shown to be of little real value.

From one and the same portion of fat dissolved in chloroform, for instance, four or five different kinds of crystals may be obtained, and by using other solvents or changing the temperature of crystallization still different forms may be secured. This fact is illustrated in the figures above which are very exact illustrations of the microscopic appearances. They are copies of microphotographs made by the author and show as observed under a magnifying power of about 200 diameters.

Chapter IV.

PROTEIDS OR ALBUMINS.

THE term proteid is applied to a large number of substances derived from different sources which are very similar in composition, and which in many instances appear to be impure forms of one and the same compound.

These proteid compounds are found in both the animal and vegetable kingdoms, but never in pure form. Several closely related varieties are often associated, and mixed with these are fats, carbohydrates, mineral salts, etc., in varying proportions.

Practically, the preparation of a pure albumin is a matter of very great difficulty, and the analyses made by different experimenters do not show remarkably close agreement. It is, therefore, impossible to assign a formula to these bodies, although this has several times been attempted. The outside limits assigned for the percentage composition are given below:

```
C............................50.0 to 55.0
H............................ 6.8  "  7.3
O............................20.0  " 24.1
N............................15.0  " 18.2
S............................ 0.3  "  5.0
```

As an illustration of the results reported by a single observer the analysis of egg albumin by *Hammarsten* may be quoted. He found for the dry substance:

```
C............................52.25
H............................ 6.90
O............................23.67
N............................15.25
S............................ 1.93
                             ──────
                             100.00
```

Albumins as a class are recognized by the following reactions:

(*a*.) They give the characteristic test for nitrogenous bodies when fused with soda-lime.

(*b*.) They yield precipitates with solutions of salts of certain heavy metals.

(*c*.) They give, especially, red precipitates when warmed with Millon's reagent.

(*d*.) They give a characteristic color when heated with strong nitric acid, and are then treated with ammonia.

(*e*.) They give color reactions when mixed with a small amount of copper sulphate solution, and then with an excess of alkali.

These reactions are illustrated by the following experiments:

Ex. 51. Mix some common wheat flour with an equal bulk of soda-lime in a test-tube, and heat strongly in the flame of a Bunsen burner. Ammoniacal vapors are given off, which may be recognized by the odor, or by action on litmus paper. The ammonia is produced in the decomposition of the proteid matter of the flour by the fixed alkali. This is a characteristic and delicate reaction.

Prepare a dilute solution of white of egg and use it for tests to follow. Pour the white of one egg into a strong bottle holding about half a liter. Add 100 Cc. of distilled water and some small quartz pebbles, or better, some pieces of clean broken glass, and shake vigorously. After a time add 400 Cc. more and shake again, and allow to stand about fifteen minutes. Strain through clean, unsized muslin, and use the nearly clear liquid for the tests.

Ex. 52. Pour some of the liquid into several test-tubes and add small amounts of solution of mercuric chloride, copper sulphate, lead acetate and silver nitrate. Note that each reagent produces a precipitate.

Ex. 53. To another portion of the liquid in a test-tube add half its volume of Millon's reagent (for preparation of this see appendix), and heat. A yellowish precipitate which appears at first becomes red on boiling.

This test shows equally well with other albumins and is very delicate. With very dilute solutions a red color appears without precipitation.

Ex. 54. Xanthoproteic reaction. To some albumin in a test-tube add strong nitric acid, and boil. A yellow precipitate or color is formed which, on addition of ammonia, becomes deep orange red.

The colors obtained in this reaction depend on the strength of the albumin solution. With weak solutions a precipitate may not form at all, but the yellow color is characteristic. When only traces of albumin are present the yellow color with the nitric acid may fail to appear, but the addition of ammonia gives the final part of the test with a yellow instead of an orange red color.

Ex. 55. To some of the albumin solution in a test-tube add enough sodium hydroxide to give a very strong alkaline reaction. Then add two or three drops of a dilute copper sulphate solution. This gives a violet color, which by application of heat becomes much deeper. This is known as the *biuret reaction*, and serves for the detection of small traces of albumins in solution. It is absolutely necessary to employ a considerable excess of the alkali with only a very small amount of copper solution.

Ex. 56. Make some albumin solution strongly acid with acetic acid, add some sodium sulphate and boil. All proteids except peptones are precipitated in this manner, and the reaction is employed to free solutions from these bodies. The filtrate, after boiling, can be used for other tests as the acid and sulphate exert no decomposing action. The method is frequently employed in urine analysis when it is desirable to test for small amounts of sugar in presence of relatively large quantities of albumin.

Classification of Albumins.

Investigation has shown that the bodies of this group fall naturally into several classes. At the present time it is usual to consider seven of these classes, the division being based on certain features easily recognized. It must be remembered, however, that this division is somewhat arbitrary and is not universally accepted as here given, according to *Hoppe-Seyler* and others.

CLASS I. Native albumins... { Egg albumin, Serum albumin.

CLASS II. Derived albumins. { Alkali albumin, Acid albumin, Casein, Syntonin.

CLASS III. Globulins......... { Crystallin, Vitellin, Paraglobulin, Fibrinogen, Myosin, Globin.

CLASS IV. Fibrin.

CLASS V. Coagulated proteids.

CLASS VI. Albumoses and peptones.

CLASS VII. Lardacein.

NATIVE ALBUMINS.

These are soluble in water, but are precipitated by boiling or by addition of alcohol.

Their solutions are not precipitated by sodium chloride or sodium carbonate.

Egg Albumin. Different kinds of eggs furnish albumins of slightly different properties, but these variations may be due to accompanying substances rather than to inherent differences in the albumins themselves. Most of our knowledge of these bodies has been derived from a study of the hen's egg, and this will be taken for the experiments below. Prepare a solution of white of egg as given above and make tests with it as follows:

Ex. 57. Note that a small portion of the solution is readily coagulated by heat in a test-tube. By the aid of a thermometer find approximately the temperature of coagulation, which is not far from $60°$ C. Some of the proteids present in white of egg begin to coagulate, it is claimed, at even a lower temperature while a temperature of $80°$ C. may be reached before coagulation is complete.

Ex. 58. Warm three or four Cc. of strong nitric acid in a test-tube and pour in about an equal volume of the albumin solution carefully, so that it will float on top and not mix with the acid. A coagulation of the albumin takes place at the juncture of the two liquids and appears as a white ring. This is a valuable and delicate test.

Ex. 59. To ten Cc. of the egg solution add powdered magnesium sulphate to saturation. A precipitate of globulin separates out. Filter and add to the filtrate some strong solution of sodium sulphate. Precipitation of the albumin takes place now, if the solution is kept at a temperature of $40°$ C. This precipitation seems to be due to the presence of a double sulphate of magnesium and sodium in the liquid.

Serum Albumin. Serum albumin occurs in the blood along with several other proteids, but can be obtained nearly pure by the following method:

Fresh blood from the slaughter house is poured into a shallow dish and whipped with a bunch of twigs or agitated vigorously with an egg beater to separate the fibrin. The latter is strained out by means of clean unsized muslin. The liquid passing through the muslin is put into a centrifugal apparatus and rotated to throw out the corpuscles. If the centrifugal apparatus is not at hand the fibrin-free liquid can be poured into a tall glass jar and allowed to stand until the corpuscles settle. The clarified liquid obtained by either method is poured into a large beaker and mixed with four times its volume of saturated solution of ammonium sulphate, or enough to yield a completely saturated liquid. This precipitates the albumin and globulin present, and these compounds are then separated by filtration. The moist mass on the filter is dissolved in a small quantity of water and treated again with an excess of ammonium sulphate which gives a purer precipitate. The product is washed on the filter with strong ammonium sulphate solution, then dissolved in a little water and subjected to dialysis, by which means the salts are eliminated while an albumin solution remains on the dialyser.

During the dialysis globulin separates and at the end of the operation is filtered out leaving a nearly pure serum albumin. This can be further purified by adding a little ammonia until the reaction becomes neutral, dialyzing again and finally filtering. The clear filtrate can be used for tests or further concentrated. To obtain it in dry form evaporate at 40° C. to a small bulk, add strong alcohol in excess to precipitate, filter without delay, press out the alcohol, and displace the remaining alcohol by washing with ether. Dry at a low temperature, powder and preserve in a well-stoppered bottle.

As shown by the method of preparation serum albumin is soluble in water, but insoluble in saturated solution of ammonium sulphate. The tests given for egg albumin are true for serum albumin. These points of difference may be noted:

Ex. 60. Note that a neutral solution of egg albumin is coagulated by ether, while the solution of serum albumin is not.

Ex. 61. Coagulate a solution of egg albumin by heat and note that the precipitate dissolves but slightly in boiling nitric acid. A precipitate of serum albumin dissolves in nitric acid.

Ex. 62. Precipitate solutions of egg albumin and serum albumin with nitric acid and note that the latter precipitate is much more readily soluble in excess than is the former.

DERIVED ALBUMINS.

These are nearly insoluble in pure water or dilute saline solutions, but are soluble in weak acids or alkalies. Boiling does not coagulate the solutions.

Alkali Albumin. This is most readily prepared by the action of alkali on native albumins.

Ex. 63. Add strong sodium hydroxide solution to white of egg, with constant stirring, until a thick jelly is formed. Too much alkali must not be added here, but just enough to make the maximum of jelly. This is now cut into small pieces and washed in distilled water several times until the lumps are white throughout. They are then heated with fresh pure water, but very gently, until they go into solution. This is then filtered and the filtrate precipitated by acetic acid, avoiding any excess. The precipitate is washed with pure water.

Observe that this substance is soluble in dilute acid or alkali solutions.

Ex. 64. Bring some of the washed alkali albumin into solution with hot water. Add a few drops of phenol phthalein solution, which imparts a red color, and then run in dilute sulphuric until the alkali reaction just disappears. As soon as the neutral point is reached a precipitate falls, but, if the addition of acid be continued it will disappear with formation of acid albumin.

The thick jelly-like substance first formed by the action of strong alkali on undiluted white of egg solution is called *Lieberkuehn's* jelly. It contains an excess of uncombined alkali. A weak solution of alkali albumin can readily be made by warming diluted white of egg solution with a small amount of very weak alkali solution. The temperature should be maintained at 35° C. to 40° C., for about an hour. At the end of this time the solution can be boiled without coagulation.

Acid Albumin. The bodies called acid albumins are very similar in properties to the alkali albumins and at one time were supposed to be identical with them. But careful experiments have disclosed several points of difference. Acid albumin is most readily made by the action of acid on native albumins.

Ex. 65. Dilute white of egg with four volumes of of water, take 25 Cc. of the mixture, add 5 Cc. of 0.2 per cent hydrochloric acid and warm it on the water-bath for about two hours to a temperature of 45°. Then carefully neutralize the solution with dilute sodium hydroxide. This precipitates insoluble acid albumin, which can be washed with water by decantation. It is essential that just the right amount of alkali be added here, an excess would redissolve the precipitated acid albumin with formation of alkali albumin. The washed acid albumin can be used for a number of tests.

Ex. 66. Redissolve a small portion of the precipitate with hydrochloric acid, and observe that the solution is not coagulated by boiling.

Mix another portion of the precipitate with water and heat to 75° or 80°. This brings about a coagulation similar to that produced with native albumins.

Acid albumins are very quickly formed by the action of strong hydrochloric or acetic acid on white of egg. Coagulated proteids may also be converted into acid albumins by a much longer digestion.

The striking peculiarity of this acid albumin is that it is insoluble in pure water, but does dissolve in weak acid or alkali. It gives the characteristic yellow color with strong nitric acid, a reddish yellow precipitate with Millon's reagent and the *biuret* reaction.

Acid albumin is less soluble than alkali albumin and appears to differ also in specific rotation.

Casein. Milk contains several proteid substances, the most abundant of which is casein. In normal cow's milk it is present to the extent of about four per cent, while in woman's milk it appears to amount to two and one-half per cent or less. While the determination of casein in cow's milk has been made with great accuracy and is an operation of no special difficulty there seem to be practical difficulties in the way of carrying out this determination in the case of human milk. The first point of uncertainty is found in the collection of a normal sample, and from this and other causes it follows that the results given by different chemists for the amount of casein present vary between 0.6 and over 5 per cent.

Equally great variations have been found in the milk furnished by several animals.

Our practical knowledge of casein has been in the main derived from a study of the product of cow's milk.

In most of its reactions casein is closely allied to alkali albumin and by some has been supposed to be identical with it. But other reactions disclose certain differences which seem to be essential and the weight of evidence appears to be against the identity of the two proteids. The important properties of casein are made apparent by the following experiments.

Ex. 67. To 100 Cc. of fresh milk add 500 or 600 Cc. of distilled water and then enough dilute acetic acid to give a distinct acid reaction. This will throw down a copious white precipitate of casein with fat which is allowed to settle and washed with distilled water by decantation. The precipitate is then transferred to a filter and allowed to drain and then washed with 50 Cc. of strong alcohol. This removes water and part of the fat present. The precipitate is finally washed with ether to remove the remainder of the fat and allowed to dry in the air.

After the above treatment the casein is left as a white powder pure enough for tests. A better product is obtained by dissolving the precipitate first thrown down in weak soda solution and passing the liquid through several filters. It is then reprecipitated with dilute acetic acid, washed by decantation, collected on a filter and finished as before.

Ex. 68. Test a small amount of the casein prepared according to the last experiment as to its solubility in water. This will be found to be very slight. It dissolves readily in weak alkali solutions.

Ex. 69. Mix 5 Cc. of fresh milk in a test-tube with 10 Cc. of saturated salt solution. Then add a little more powdered salt, close the test-tube with the thumb and shake vigorously. Casein with fat is precipitated and can be collected on a filter. Boil the opalescent filtrate and note the coagulation of a soluble albumin present. In cow's milk this form of albumin amounts to about 0.5 per cent.

An excess of magnesium sulphate solution may be used instead of the sodium chloride to precipitate the casein.

Under the action of rennin, casein undergoes a peculiar modification resulting in the formation of a clot or curd. The nature of this will be explained in the chapter on milk.

The behavior of milk with rennin serves to distinguish casein from true alkali albumin. With the latter, it is said, no clot can be obtained.

Pure casein yields no ash on ignition, although it contains phosphorus and sulphur in organic combination. Pure alkali albumin yields a small amount of ash when ignited.

Syntonin. This appears to be an acid albumin, resulting from the action of dilute acids on muscle.

Ex. 70. Free the muscular part of meat from fat as far as possible and run it through a sausage mill several times to bring it to a fine state of subdivision. Wash this chopped mass with distilled water until the washings remain clear. Now, to about 5 Gm. of the moist residue in a small flask, add 50 Cc. of dilute hydrochloric acid, containing $\frac{1}{10}$ per cent of the true acid. Warm the mixture slightly (to 35° or 40° C.), and keep at this temperature about three hours. Then filter and test the filtrate. It contains the soluble syntonin.

Ex. 71. To a small portion of the filtrate, add weak caustic soda, which produces a precipitate soluble in excess of the alkali.

Boil another portion of the filtrate. It does not coagulate.

Globulins.

We have here a peculiar class of proteid bodies, insoluble in pure water and strong saline solutions, but soluble in weak saline solutions. A globulin dissolved in a weak solution of salt is precipitated if the same salt is added to saturation. Magnesium sulphate is most active in producing this precipitation. The globulins resemble the native albumins in being coagulable by heat; they are also insoluble in alcohol. Dilute acids or alkalies dissolve them by forming acid or alkali albumin.

Crystallin. This is the globulin of the crystalline lens, and can be obtained by extracting the latter, in finely divided form, with a one per cent salt solution.

Vitellin. Yolk of egg contains a large amount of this globulin, associated with other substances of which lecithin is, perhaps, the most important. The preparation of pure vitellin is a matter of great difficulty because of the presence of the lecithin which cannot be readily separated. A partial separation can be effected by extracting with ether, leaving a residue which dissolves easily in salt solution. A great excess of sodium chloride does not reprecipitate, as is the case with other globulins.

Paraglobulin, or serumglobulin occurs with pure albumin in blood serum, from which it can be separated in nearly pure condition.

It was shown above, under the head of blood serum, that in the process of dialysis for the final purification of the latter, globulin separates out as the saline liquid loses its salts and becomes aqueous. The precipitation of globulin can be illustrated by the following experiment:

Ex. 72. Separate fibrin and corpuscles from blood as described above under serum albumin and saturate the resulting liquid with magnesium sulphate. This throws down a precipitate of globulin, which can be washed on a filter with a saturated solution of the sulphate without much loss. By mixing the precipitate with a little water, and gradually adding more, it will finally dissolve, but the addition of a very large volume of water produces a precipitate again.

Common salt, added in excess, acts as does the magnesium sulphate.

The method illustrated by the above experiment is not sufficient for the preparation of pure globulin. A better product is obtained by dissolving and reprecipitating several times with magnesium sulphate. Finally the precipitate may be dissolved in the right amount of water and dialyzed for the separation of the salts. The globulin remains on the dialyzer as an insoluble residue.

Fibrinogen. In the clotting of blood this substance plays an important part as in the operation it becomes converted into the stringy fibrin which encloses the corpuscles forming a thick and jelly-like mass.

Myosin. This is a globulin formed in muscle after death and is produced from a substance resembling fibrinogen existing in the living muscle. The name myosinogen has been given to this primary form.

The preparation of myosin is illustrated by the next experiment:

Ex. 73. Free muscle (round steak) as far as possible from traces of fat and sinews and then thoroughly disintegrate it by passing through a sausage mill. Then wash it repeatedly with cold water until the latter is no longer reddened, and the residue appears white. This is placed in a 20 per cent solution of ammonium chloride and allowed to remain about a day, with occasional shaking. Myosin dissolves in the ammonium chloride and is found in the filtrate when the mixture is filtered. Pour the filtrate into twenty times its volume of distilled water, which causes a precipitation of the insoluble myosin. Allow to settle and wash three times by decantation. Collect the precipitate and observe that portions of it dissolve readily in ten per cent solutions of sodium chloride and ammonium chloride or in a 0.1 per cent solution of hydrochloric acid. The solution in salt is precipitated by the addition of more to saturation.

Myosin in salt solution coagulates at a low temperature, usually given as 55° to 56° C. By acids it is easily converted into syntonin.

Globin is sometimes described as a distinct globulin, formed by the spontaneous decomposition of hæmoglobin, but no great amount of study has been given to it by chemists.

Fibrin.

Fibrin is the name commonly given to the substance obtained by whipping blood with a bundle of twigs. So prepared it is very impure, holding white and also some red corpuscles. By other means, however, a much purer product can be obtained. Made in any manner fibrin is a white elastic substance insoluble in water and weak salt solutions. It dissolves gradually in acid or alkali solutions, forming proteids of other groups. A corresponding vegetable product can be easily made by mixing flour with a little water to form a stiff dough. After standing some time this is worked with the hand under running water which washes out the starch, leaving finally the vegetable fibrin or gluten as a white stringy mass.

Animal fibrin is produced from the proteid referred to above as fibrinogen, occurring in the blood before death.

Coagulated Proteids.

We have here a class of substances produced by action of heat on most of the proteid bodies just described. When a proteid assumes the coagulated form it becomes insoluble in water, dilute acids or alkalies, and can be dissolved by strong acids only through complete modification and partial destruction. Alcohol has also the property of coagulating many proteid substances.

We have a familiar illustration 'of a coagulated proteid in boiled white of egg. This, and similar substances, undergo a peculiar change called digestion when taken into the stomach where they are acted on by the gastric juice, or in the intestines under the action of the pancreatic juice. This digestion will be considered in a few experiments later, as it is one of the important reactions taking place in the animal body.

ALBUMOSES AND PEPTONES.

Under the action of the gastric juice and the secretion of the pancreas just referred to albuminous substances in general suffer a profound modification whereby they become soluble and ready for absorption as preliminary to nutrition. The gastric juice is a weak acid liquid of rather complex composition, but containing at least two substances of importance which merit our attention here.

The first of these is a peculiar enzyme or soluble ferment called pepsin and the second is hydrochloric acid. This acid is present in all normal gastric juice and in man it amounts to about 0.25 per cent of the total weight of the juice. The other substances present have as a matter of course some important function, but the manner of behavior of the acid and pepsin have been made the subject of many investigations. The essential work of these two compounds is to convert other proteids into the forms known as albumoses and peptones, of which the latter are extremely soluble and to some degree diffusible. Between the ordinary coagulated proteids taken into the stomach as food and the final peptone or completely digested matter there are possibly a number of intermediate stages. The albumoses represent one of these and can be distinguished from the end product by several fairly definite reactions. It appears, further, that at the beginning of the process of digestion with pepsin and hydrochloric acid that a body similar to or identical with acid albumin is formed, to be soon followed by albumose.

In studying the digestion of starches it was shown that the pancreatic fluid contains an enzyme very active in this direction, and further that in the emulsification of fats, as a step in digestion, the pancreatic juice plays, likewise, a prominent part. The secretion of the pancreas is active finally in a third direction, viz.: in the digestion of proteids when it continues and completes the work begun by the gastric juice. Of the actual nature of the pancreatic enzymes which bring about this result little is known, but the end seems to be reached in several stages as with the gastric secretion. The first step is, apparently, the production of an alkali albumin, followed by what is termed hemialbumose as distinguished from antialbumose, formed in gastric digestion. Hemialbumose is the forerunner of hemipeptone, the product corresponding to antipeptone produced by pepsin.

There is not, however, a complete parallelism between the actions of the two digestive fluids. The work of the gastric juice ends with the production of peptone, while that of the pancreatic secretion includes the formation of several decomposition products, as leucine, tyrosine and others. These changes can be represented as follows diagrammatically (according to *Kuehne*):

Albumin.
- Antialbumose. — Antipeptone — By peptic digestion.
- Hemialbumose. — Hemipeptone — Leucine. Tyrosine. etc. — By Pancreatic digestion.

DIGESTION OF PROTEIDS.

The changes referred to above can, in part, be represented by simple experiments, of which a number will here follow.

Peptic Digestion. An active peptic ferment fluid can be readily prepared as follows:

Separate the mucous membrane of the hog's stomach from the outer coatings, and cut it into very small bits or disintegrate thoroughly in a sausage mill. Place the chopped mass in a flask or bottle and cover it with about twice its weight of good, strong glycerol. Allow to stand a week or ten days, with frequent shaking, and then pour off the glycerol, which is now a peptic extract.

For the experiments given below a good solution may be prepared by treating 5 Gm. of the minced membrane with 8 to 10 Cc. of glycerol.

Ex. 74. Boil an egg until it is hard, take out the white portion and rub it through a clean wire sieve with fine meshes, by means of a spatula. Add about 5 Gm. of this egg to 100 Cc. of 0.2 per cent hydrochloric acid in a flask, and then add 2 Cc. of the glycerol extract. Keep the flask at a temperature of 40° C., with frequent shaking. In time the egg albumin will dissolve, forming an opalescent liquid. Unless the flask is very frequently shaken the solution of the albumin will be slow. Use the solution for experiment 76.

Ex. 75. To 2 Cc. of the glycerol extract in a test-tube add a little water and boil a few minutes. Now add this boiled liquid to albumin and 0.2 per cent hydrochloric acid as in the last experiment and note that under the same conditions digestion does not take place, the heating having destroyed the active enzyme.

Ex. 76. Test for Albumose and Peptone. About half an hour after the beginning of the digestion explained in the above experiments pour off about half the liquid, neutralize it exactly with ammonia and then saturate with ammonium sulphate. This precipitates albumose but not peptone. Filter off the precipitate and apply the biuret test to a portion of the filtrate, using only a very small trace of copper sulphate. A pink color should be observed. Concentrate the remainder of the filtrate from the albumose precipitate by evaporation at a low temperature, pour it in a test-tube and add some strong alcohol. This gives a precipitate of peptone, which dissolves by adding water.

The half of the liquid of Ex. 74, not used in the last test, if allowed to stand long enough will give tests for peptones only, the albumose being entirely converted. It will afford the student good practice to carry out experiments on the digestion of albumin, using some of the prepared pepsins on the market. These products are made by several processes from the hog's stomach and vary greatly in their digestive power.

Comparisons between different commercial pepsins must be made by some arbitrary method, but carefully conducted as to temperature, amount of liquid employed, character and condition of the albumin taken for experiment, and so on. A process frequently followed is here given:

PROCESS OF THE UNITED STATES PHARMACOPŒIA FOR VALUATION OF PEPSIN. 1890.

"Prepare, first, the following three solutions:

A. To 294 Cc. of water add 6 Cc. of diluted hydrochloric acid, ten per cent.

B. In 100 Cc. of solution A dissolve 0.067 Gm. of the pepsin to be tested.

C. To 95 Cc. of solution A, brought to a temperature of 40° C., add 5 Cc. of solution B.

The resulting 100 Cc. of liquid will contain 0.2 Cc. (0.21 Gm.) of absolute hydrochloric acid and 0.00335 Gm. of the pepsin to be tested.

Immerse and keep a fresh hen's egg during fifteen minutes in boiling water; then remove it and place it in cold water. When it is cold, separate the white coagulated albumin and rub it through a clean sieve, having thirty meshes to the linear inch. Reject the first portion passing through the sieve. Weigh off 10 Gm. of the second, cleaner portion, place it in a flask of the capacity of about 200 Cc., then add one-half of the solution C and shake well, so as to distribute the coherent albumin evenly throughout the liquid. Then add the second half of the solution C, and shake again, guarding against loss. Place the flask in a water-bath, or thermostat, keep at a temperature of $38°$ to $40°$ C., for six hours, and shake it gently every fifteen minutes. At the end of this time the albumin should have disappeared, leaving at most only a few thin insoluble flakes. (Trustworthy results, particularly in comparative trials, will be obtained only if the temperature be strictly maintained between the prescribed limits, and if the contents of the flasks be agitated uniformly and in equal intervals of time.)

The relative proteolytic power of pepsin, stronger or weaker than that described above, may be determined by ascertaining, through repeated trials, how much of solution B made up to 100 Cc. with solution A will be required exactly to dissolve 10 Gm. of coagulated and disintegrated albumin under the conditions given above."

The pharmacopœia of the United States requires that pepsin must be able to digest 3,000 times its weight of coagulated albumin and the above process serves to determine whether or not a given sample comes up to the accepted standard. Another method depending on the same general principles may be described as follows:

General Process.

Prepare a weak acid and pepsin solution as described above and determine by two or three trials the weight of albumin it will digest. Or, most conveniently, to five portions of 100 Cc. each of the acid-pepsin mixture add 6, 8, 10, 12 and 14 Gm. of the prepared egg albumin and digest in flasks holding about 200 Cc. The flasks must be immersed in a large water-bath, the temperature of which is kept uniform by frequent stirring, and must be agitated at regular intervals and to the same extent through five hours. This length of time is sufficient to show approximately the digesting value of the pepsin.

Now weigh out accurately an amount of the prepared egg albumin about one-third greater than the largest amount dissolved in the flask experiments above, and treat with 100 Cc. of the acid-pepsin mixture in the same manner. At the end of six hours a part of this albumin will be still undissolved, and must be filtered off, dried and weighed. The weight subtracted from that taken, gives the amount which has been actually digested.

Some precautions are necessary in the filtration and weighing. In order to obtain a liquid which can be filtered readily, it has been recommended to add to it an accurately weighed amount (about 3 Gm.) of finely divided, well washed and dried asbestos, such as is used in the Gooch funnels, with 100 Cc. of water, and shake thoroughly. The mixture is then poured on a large weighed Gooch funnel, washed thoroughly with distilled water, and finally with some alcohol. It is then dried at 110° C. to a constant weight and accurately weighed. From the gross weight of the contents subtract the weight of the asbestos. The remainder represents the weight of the *dry* albumin. Multiply this by 7.5, to obtain the moist equivalent, and subtract this from the weight originally taken, to find that actually digested.

In the digestion experiments given above, it has been directed to use weak hydrochloric acid with the active enzyme. This is necessary as can be shown by a trial.

Ex. 77. Pour about 50 Cc. of 0.2 per cent hydrochloric acid into a flask, add about half a Cc. of the glycerol extract of pepsin and a gram of finely divided hard boiled white of egg. In a similar flask take 50 Cc. of distilled water with the same amounts of pepsin extract and albumin as before. Place both flasks in water at a temperature of 40° C., and keep them there about an hour. In the flask to which the hydrochloric acid had been added, the digestion will be found far advanced, or complete, while in the other no change will be observed.

Test the liquid of the first flask for peptones.

Lactic acid is frequently found in the gastric juice in small amount, but probably as a product rather than as a factor in peptic digestion. Hydrochloric acid is always normally present, but under pathological conditions the amount may be very much diminished, or nearly absent. It becomes, at times, a question of importance to determine the actual amount of this acid present in the juice as collected by a tube. Several tests suffice for the recognition of hydrochloric acid in presence of lactic or other organic acid, some account of which will be given in a subsequent chapter.

Pancreatic Digestion of Proteids. Under the head of digestion of starch directions were given for the preparation of a pancreatic extract which serves, also, very well for the digestion of proteids. It was shown that the active ferment of the gastric juice acts only in an acid medium; the ferment of the pancreas acts in a solution which is alkaline in reaction. The mode of action of the pancreatic enzyme can be shown by the following experiment

Ex. 78. Pour 25 Cc. of a 1 per cent solution of sodium carbonate (crystallized salt) into each of several small flasks or test-tubes. Add to each half a Cc. of the glycerol extract of pancreas and about a gram of finely divided hard boiled white of egg. (The white of egg can be easily prepared according to the methods given above under the head of pepsin testing.) Make one of the tubes slightly acid by the addition of dilute hydrochloric acid. Now place all of them in water kept at 40° C. At the end of half an hour remove one of the alkaline tubes, and note that it still contains unaltered coagulated albumin. Test the liquid for albumoses and peptones as given above, Ex. 76. After another half hour, test a second tube (after filtration). It will be observed that as the coagulated proteid disappears, peptones become more abundant.

Allow one of the alkaline tubes to remain several hours at a temperature of 40° C. In time it develops a disagreeable odor, due to the presence of indol formed. The tube containing the hydrochloric acid kept several hours at 40° C. does not show the effects of digestion, indicating that an acid medium does not suffice for the converting activity of the pancreatic ferment.

To readily recognize the final products of the pancreatic digestion of proteids it is necessary to start with larger quantities of materials than are given in the above experiment. By the following method enough of the products for demonstration in a class can be secured:

Ex. 79. Mix 25 Gm. of fresh minced pancreas with an equal weight of fresh fibrin and 250 Cc. of thymolized water, in a flask. Place the flask in a thermostat and keep it at a temperature of 40° C. six hours. At the end of this time pour off about one-third of the mixture, boil it and filter warm. Concentrate the filtrate to a small bulk and place drops on glass slides for further evaporation preliminary to microscopic examination. Leucine, one of the products, crystallizes in spherical bunches of minute short needles, while tyrosine, another, crystallizes in long needles often bunched together. Tyrosine, even in small quantity gives a rose red color when boiled with Millon's reagent.

The larger portion of the digesting mixture is allowed

to remain six hours longer in the thermostat at 40° C., for the production of indol the presence of which is shown by the characteristic odor of the fluid and also by chemical tests.

A splinter of pine wood moistened with hydrochloric acid turns red when dipped in a small portion of the liquid previously acidified with strong hydrochloric acid.

A red color is also imparted by the action of a small amount of nitrous acid on indol. To obtain this test use 25 Cc. of the liquid, neutralize the alkali by addition of dilute sulphuric acid, and leave a *very little* acid in excess. Then add a small amount of a solution of sodium or potassium nitrite, which brings out the reaction.

In the above experiments leucine and tyrosine appear to be true products of pancreatic digestion, while indol is formed by a subsequent putrefactive process. Leucine and tyrosine are very common and frequent decomposition products formed from proteids, or allied compounds, by a variety of reactions. In a following chapter a process will be given by which they may be obtained through the action of weak acids on horn shavings.

Diffusion of Peptones. Peptones are soluble in water, and diffusible, but their rate of diffusion is so slow that it cannot be readily observed in a simple laboratory experiment. It has commonly been held that this diffusion plays a very important part in the absorption of the digestive products from the intestines, but the phenomenon is of so complex a nature that the part played in it by diffusion or osmosis is one which cannot be clearly defined.

Lardacein or Amyloid Substance.

This is a pathological product closely related to the proteids, found in several organs of the body.

Most authorities agree that it cannot be digested by the peptic or pancreatic enzymes in which respect it differs from the common proteids, but it resembles them in yielding acid albumin with strong acids, and alkali albumin on treatment with alkali hydroxides.

Lardacein is usually separated from finely divided tissue containing it, by washing out everything soluble in water and dilute alcohol and then digesting with pepsin and hydrochloric acid. The lardacein is left, with small amounts of other substances, as an insoluble residue.

It gives the color reaction with Millon's solution and is specially characterized by its behavior with several reagents, giving a red or brown color with iodine and a violet to pure blue with iodine and sulphuric acid. Hence the name, from fancied resemblance to amylum. It gives color reaction with several of the aniline colors also.

Chapter V.

THE BLOOD.

THE blood of man, normally, is an opaque viscid fluid, red in color and alkaline in reaction. It is not and cannot be constant in composition, because of the several functions it is called upon to perform. The blood carries nutrient matter to the tissues and must vary in density and complexity according to the nature and amount of the nutritive supply. It carries away waste matter from the tissues and delivers it at the principal points of elimination. The character of the blood must depend on the rapidity of this elimination.

Blood contains a characteristic coloring matter which has the peculiar property of combining with oxygen and other gases with variations in color. Oxidized blood is bright red, while reduced blood or blood combined with other gases than oxygen is much darker in hue.

The general composition of blood may be represented by this scheme.

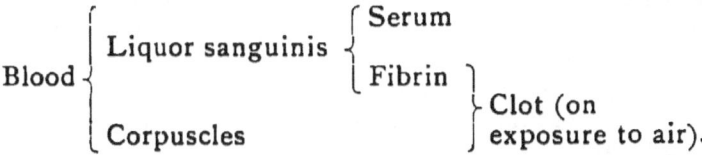

On coagulation the corpuscles become entangled with the stringy fibrin, forming what is commonly termed the clot.

The percentage composition of average normal human blood has been determined by several chemists. The following results (*Becquerel* and *Rodier*) are frequently quoted.

IN 1,000 PARTS BY WEIGHT.

Water..779
Solids...221
Fibrin... 2.2
Hæmoglobin..134.5
Albumin... 76.0
Cholesterin, fat and lecithin................... 1.6
Salts and extractive matters................... 6.8

General Tests for Blood.

Blood may be recognized by its appearance under the microscope, in which case the corpuscles are characteristic, by the absorption bands of its solution when viewed through a spectroscope and finally by certain chemical tests. Some of these will be given here by way of illustration. Use blood diluted with about twenty volumes of water for the tests.

Ex. 80. Heat the solution of blood until it is near the boiling temperature and note that the red color is largely destroyed and that a brownish precipitate forms which contains albumin and decomposed coloring matter. Add now a small amount of sodium hydroxide solution and observe that the precipitate disappears while the blood solution becomes red again by reflected light, but greenish by transmitted light.

Ex. 81. Guaiacum Test. To a little blood solution in a test-tube add some *fresh* tincture of guaiacum and then a few drops of an ethereal solution of hydrogen peroxide. Shake the mixture and observe that the precipitated resin has assumed a blue color, more or less marked. In this test turpentine oil which has been exposed to the air, or which has been shaken with air in a bottle, can be used instead of the solution of peroxide. Hydrogen peroxide is developed by the action of oxygen on turpentine.

This test depends on the oxidation of the resin by the peroxide in presence of blood and has practical applications.

Hæmin Crystals. When acted on by acids or strong alkalies the hæmoglobin of blood is broken up into *globin* and a characteristic compound called *hæmatin*. Hæmatin in turn is decomposed by hydrochloric acid yielding *hæmin* which appears in crystalline form. From the name of their discoverer, these crystals are called Teichmann's crystals. Their appearance constitutes one of the best tests we have for blood, and can be illustrated by the following:

Ex. 82. Evaporate a drop of blood on a slide, add two or three drops of glacial acetic acid and boil. Put on a cover glass and allow to cool. Minute (microscopic) plates or prisms separate out. If old blood, a stain for instance, is examined, it is necessary to add a small crystal of sodium chloride to the acetic acid, by which means sufficient hydrochloric acid is liberated for the test.

The crystals have a dark brown color and are very characteristic.

Reaction of Blood. The reaction of normal fresh blood is alkaline, which can be shown as follows:

Ex. 83. Prepare some small smooth plaster of Paris surfaces by pouring the well-known plastic mixture of plaster of Paris and water on glass plates and allowing it to harden several hours at least. The prepared plates are removed from the glass and soaked in a *neutral* solution of litmus and are then allowed to dry. The test proper can now be made by putting a few drops of the blood on the smooth plaster surface and allowing it to remain there five minutes. It is then washed off with pure water when it will be found that the part of the plate which had been covered by the blood has become blue from the action of the alkali of the blood on the neutral litmus.

The Coagulation of Blood. Some of the simpler phenomena connected with the coagulation of blood may be readily shown by experiment.

Ex. 84. Have ready two test-tubes. Pour into the first one Cc. of a cold saturated solution of sodium sulphate, the other is left clean and dry. Decapitate a rat and allow two Cc. of the escaping blood to flow into the tube containing the sodium sulphate. The rest of the blood is collected in the dry tube. In a very few minutes coagulation takes place in the latter tube, while it is prevented by the sodium sulphate in the former.

Allow both tubes to stand at rest a day or more. In the salted tube it will be noticed that most of the corpuscles have settled to the bottom, leaving a clear and lighter colored liquid, while in the other tube the coagulum has begun to shrink into a smaller mass from which droplets of yellowish serum ooze. The corpuscles in this case remain with the fibrin.

Ex. 85. Collect a quantity of slaughter house blood by running two volumes of the latter into one volume of saturated solution of sodium sulphate. Shake the mixture and allow it to stand at a low temperature several days. Coagulation does not occur, but a gradual precipitation of the corpuscles is observed, leaving a yellowish liquid known as salted plasma, which may be poured off and used for various experiments.

Ex. 86. Pour a few Cc. of the salted plasma into a test-tube and dilute it with several times its volume of water. On slight warming of the mixture coagulation follows. The effect of the sodium sulphate is to prevent coagulation. In this case dilution favors it.

Hæmoglobin.

This is a complex proteid-like body which can be obtained in crystalline form, but which is not diffusible. It makes up the larger part of the solid matter in the red corpuscles, and is distinguished from the true proteids by containing a small amount of iron.

In the experiments given above the hæmoglobin remains with the corpuscles, and as they settle out the serum is left as a yellowish, clear liquid. If from any cause the corpuscles become disintegrated the hæmoglobin separates, allowing the residue to settle, and imparts now a permanent red color to the serum. This disintegration of the corpuscles can be effected in several ways, by addition of water to blood, by repeated freezing and thawing, by treatment with certain salts, and by shaking with ether.

In normal arterial blood hæmoglobin exists in the oxidized form known as oxyhæmoglobin; in venous blood it exists in a partly reduced condition. The most characteristic property of hæmoglobin, and the one on which its value in the blood mainly depends is its power of combining with certain gases to form loose compounds easily broken up. Oxyhæmoglobin is such a compound, and by means of it oxygen is conveyed through the blood from the lungs to the remote tissues. The oxygen there being given up as required, the blood returns by the venous system, darker in color, and containing "reduced" hæmoglobin, or hæmoglobin proper.

The action of oxidizing and reducing agents on blood can be illustrated by a few simple experiments.

Ex. 87. Shake about 10 Cc. of defibrinated blood with a few drops of ammonium sulphide solution or with Stokes' reagent. (This is a solution of ferrous sulphate, to which a small amount of tartaric acid has been added, and then ammonia enough to give an alkaline reaction). Warm gently, and observe that the bright color of arterial blood gives place to the darker purple of venous. On shaking the mixture now with air the bright red color returns. For the success of this experiment where Stokes' reagent is employed it should be freshly prepared before use. Various other substances behave in a similar manner.

Ex. 88. Generate some hydrogen gas in the usual manner, and allow it to bubble through defibrinated blood. A change of color follows after a time, due to the mechanical loss of oxygen. The same result may be accomplished by exhausting the oxygen of the blood by means of an air pump. Exposure to the air restores the color in a short time, as before.

The above experiments show that oxygen can easily be removed from normal blood, and as easily restored. Hæmoglobin enters into another combination, however, which is far more stable than that with oxygen, and which, when once formed, cannot be broken up by agitation of its solution with air. This is the compound known as carbon monoxide hæmoglobin, the formation of which can be shown as follows:

Ex. 89. Pour about 10 Cc. of defibrinated blood into a test-tube, and allow a current of common illuminating gas to bubble through the liquid a few seconds. Close the test-tube with the thumb and shake thoroughly, then allow the gas to bubble through again. This should produce a change of color, the blood becoming a peculiar cherry red. Now try to restore the bright arterial red by shaking the tube in contact with air. The darker color persists, as carbon monoxide forms a very stable compound with the hæmoglobin.

This reaction is very interesting, as illustrating the action of carbon monoxide in cases of illuminating gas poisoning. Illuminating gases contain from eight to thirty per cent of this compound, and when inhaled in sufficient quantity produce death, because of a fixation of the hæmoglobin, or oxygen-carrying constituent of the blood. Blood, exposed to illuminating gas long enough, loses the power of conveying oxygen to the tissues from the lungs, and this transfer of oxygen being necessary to the maintenance of life, death must follow. Blood containing carbon monoxide can be recognized readily by the spectroscope and also by chemical tests, of which the following is one of the most reliable :

Ex. 90. Add some strong solution of sodium hydroxide to ordinary blood. This gives a *brownish green precipitate* at first and then a red solution. Treat blood saturated with carbon monoxide in the same manner. This gives a *red precipitate* and finally a red solution.

Since the oxygen carrying power, and therefore the practical value of blood, depends largely on the amount of hæmoglobin it may contain various methods have been suggested for the rapid and accurate determination of the amount of this substance present. These methods are both physical and chemical. Many analyses have been made of hæmoglobin which show, in the mean, that it contains 0.45 per cent of iron. A determination of iron gives, therefore, a measure of the hæmoglobin in a given sample of blood, and this method has been very frequently employed.

Methods depending on the comparison of the color of the blood with the color of a solution of hæmoglobin of known strength are very commonly applied at the present time and yield results sufficiently accurate for all practical purposes. Colors can be compared only in dilute solutions and for the purpose tubes similar to those used for the Nessler test in water analysis may be employed. If we have blood which may be considered normal in every way this may be taken as the standard and the sample in question compared with it. 5 Cc. of the normal blood, defibrinated, should be diluted with water to make 100 Cc. The blood to be tested is then to be diluted with water until it gives the same color as the standard when viewed under similar conditions. If observed in shallow cells a dilution with 5 Cc. of blood to the 100 Cc. is proper, but if the samples are to be compared in Nessler tubes or deep cells a dilution of 1 Cc. of blood with 99 of water will be necessary. Much better results than can be obtained by the simple Nessler tubes are given by the use of certain special instruments, such as the *hæmoglobinometer* of *Gowers* and the colorimeter, for general purposes, of *Wolff*.

In the Gowers instrument, which is shown in the cut below, a one per cent solution of normal blood is taken as

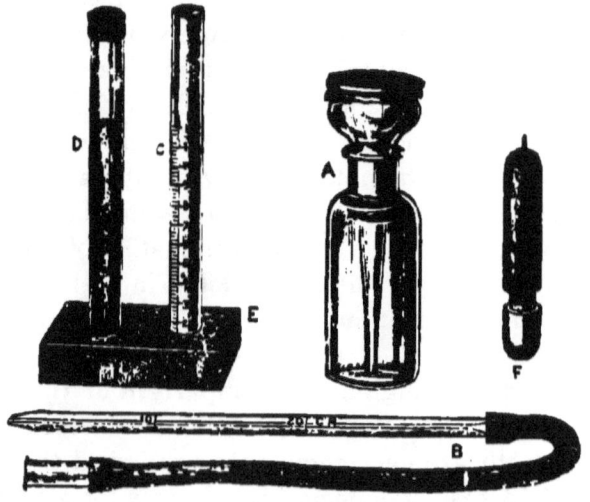

FIG. 29.

the standard and is poured into the ungraduated tube up to the mark D. Then twenty cubic millimeters of the blood under test is poured into the graduated tube and diluted with pure distilled water and shaken until the tint reached corresponds to that of the standard when viewed horizontally, the tubes having exactly the same diameter. The right hand tube shown in the figure is divided into 100 parts, or degrees, and a blood which can be diluted to 100°

before the color of the standard is reached is normal or of full strength. If the shade of the standard is reached by diluting the twenty cubic millimeters of blood to 50° its strength in hæmoglobin can be only fifty per cent of the normal amount.

Instead of using normal blood for comparison glycerol jelly, colored with picrocarmine to the proper tint, may be employed and used continuously. This is done to avoid the preparation of a normal blood solution, which must be made fresh from day to day, as it does not keep well.

The instrument does not give scientifically accurate results, but is convenient for clinical purposes and is sufficiently exact.

The cut shows a small pipette for measuring the blood, a bottle for distilled water and a puncturing needle.

Somewhat more accurate results can be secured by use of the *Fleischl* instrument or *hæmometer*, the characteristic features of which are shown in the next cut.

The standard of comparison here is a wedge of colored glass which can be moved under the transparent bottom of the observation tube and give the effect of change of shade by change in the thickness of the glass.

On the platform M is a cylindrical cell G divided into two equal parts by a vertical partition. One-half is over the movable wedge and is filled with distilled water. In the other half the diluted blood for examination is poured. By means of a screw, R, the wedge is now moved until the colors on the two sides of the partition appear the same to the eye held vertically over the cylinder. Beneath the cylinder is a white reflecting surface, S, by means of which light can be thrown from a lamp, or candle, upward through the blood and colored glass. The wedge is graduated into degrees, empirically, which indicate, usually, percentages of the normal amount of hæmoglobin in blood.

In making the test a definite, measured small quantity of blood is taken and this diluted with water up to a mark on the cylinder. The pure water in the other half of the cylinder is poured in to the same level.

In order to obtain uniform and fairly accurate results here it is necessary to measure the blood very carefully to begin with and add water then to the proper level in the two compartments. The practical value of the instrument

FIG. 30.

depends largely on the accuracy with which the normal blood colors have been duplicated in coloring the glass used. The instruments made by *Kruess*, of Hamburg, and *Reichert*, of Vienna, have been generally satisfactory.

Ex. 91. Let the student carry out practical measurements with the Gowers and Fleischl instruments, taking the blood for the purpose from his own finger by means of the shielded needle furnished with each apparatus. The amount of blood diluted must be accurately measured by the small pipette likewise furnished with the instruments.

Crystals of oxyhæmoglobin can be obtained by the following method :

Ex. 92. Mix 15 Cc. of defibrinated blood with 1 Cc. of strong ether in a test-tube. Shake thoroughly and allow to stand a day or more, the tube being stoppered. On carefully decanting the contents of the tube minute crystals will be found in the bottom which may be recognized by transferring to a slide and examining this with the microscope. These crystals differ in form according to the source of the blood; from human blood they are rhombic prisms.

A test can also be made by mixing a drop of blood (defibrinated) with a drop of water on a slide. A cover glass is put on the mixture, which is allowed to stand ten or fifteen minutes; with a microscope crystals can usually be seen at the end of this time.

The Use of the Spectroscope.

Results of great value in the examination of blood are obtained by the use of the spectroscope, with the general construction of which the student is assumed to be familiar.

When sunlight or the bright light of a lamp is focussed on the slit of a spectroscope the ordinary continuous spectrum is observed, the bright colors being interspersed, in the case of the solar spectrum, with the dark Fraunhofer lines. If, before reaching the slit, the light is made to pass through a dilute colored solution contained in a vessel with thin, clear glass walls, preferably parallel, a very different appearance is noted. The continuous spectrum is broken and perhaps wholly obscured by dark bands due to the absorption of parts of the white light by the coloring matters in the solution. The position and extent of these bands vary with the nature of the substance and the strength of its solution. Under a definite and constant degree of dilution the bands are characteristic for different substances. The spectra of oxyhæmoglobin, reduced hæmoglobin and carbon monoxide hæmoglobin have been accurately studied and constitute excellent tests for blood. The positions of these absorption bands is most conveniently represented with reference to the Fraunhofer lines which correspond to colors of perfectly definite wave length.

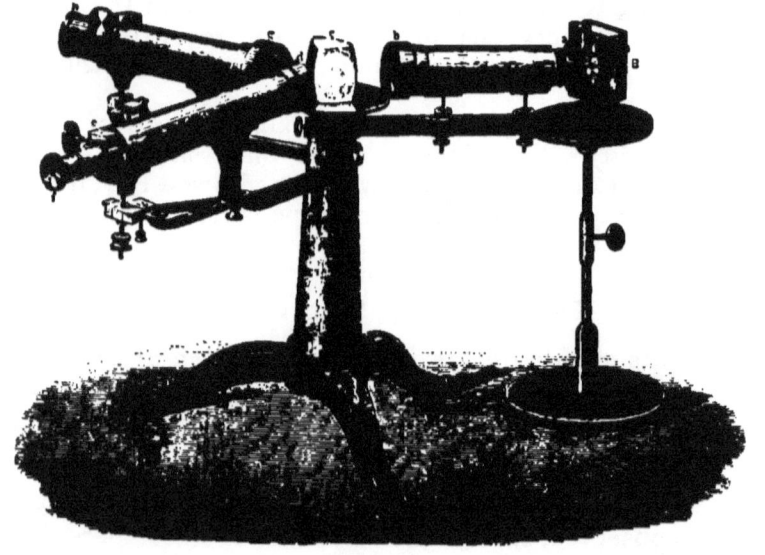

FIG. 31.

Spectroscopes to be used for anything more than rough qualitative tests must be furnished with some appliance by which the exact position of the bands in the spectrum can be determined. This is usually accomplished by the addition of a third tube to the two essential tubes, the third one containing a fine photographed scale and focussing lens so adjusted that lamp light can be thrown through the scale slit on the prism and from this reflected to the eye of the observer. The scale tube is then moved until one of its numbers is made to coincide with one of the Fraunhofer lines. Usually the line 50 of the scale is made to coincide with the D line (sodium). The position of other prominent lines is then determined with reference to the scale. The value of the scale divisions being found, once for all, the exact boundaries of any absorption band may be recorded by reference to the scale. The absorption spectrum and reflected image of the scale reach the eye at the same time. The cut opposite represents a spectroscope with absorption cell arranged for such observations.

Ex. 93. Let the student observe the absorption spectrum of oxyhæmoglobin under the following conditions: Measure accurately 5 cubic centimeters of blood and dilute it with 120 cubic centimeters of water. Mark this mixture "*Solution No. 1.*" Dilute 50 Cc. of No. 1 with 50 Cc. of water and mark the mixture "*Solution No. 2.*" Dilute 50 Cc. of No. 2 with 50 Cc. of water and mark the mixture "*Solution No. 3.*" Dilute 50 Cc. of No. 3 with 50 Cc. of water and mark the mixture "*Solution No. 4.*" Dilute 50 Cc. of No. 4 with 50 Cc. of water and mark the mixture " *Solution No. 5.*" Dilute 50 Cc. of No. 5 with 50 Cc. of water and mark the mixture "*Solution No. 6.*" Finally, dilute 50 Cc. of No. 6 with 50 Cc. of water and mark the mixture "*Solution No. 7.*"

We have now dilutions beginning with 1 in 25 and ending with 1 in 1,600. The last solution is almost colorless.

Take seven test-tubes of thin colorless glass and as uniform as possible in diameter. Number them 1 to 7 and two-thirds fill each one with the dilute blood solution corresponding to its number. Place each tube before the narrow slit of the spectroscope and adjust the flame of an oil or gas lamp so that its light may pass through the solution into the slit. Pull out the draw tube until the light is properly focussed and observe that the bright field is traversed by two black bands which cut out portions of the yellow and green. With strong blood solutions all light except the red is shut out, but with solutions of the dilutions 2 to 7 the field is obscured only by the two bands. In solution No. 2 they are very dark and well defined. With increasing dilution they grow fainter and are scarcely visible in solution No. 7. In all the solutions examined note the position of these bands with reference to the characteristic colors.

The two bands are always found in solutions of oxyhæmoglobin, and although altered in depth of shade they are not altered in position by dilution. Other red liquids may give dark bands, but blood or a solution of oxyhæmoglobin is the only liquid which gives two bands *exactly* in the position of these. We have, therefore, here a valuable means for the identification of blood and one which is very frequently applied in medico-legal investigations.

The two bands lie between the Fraunhofer lines D and E. The one near E is somewhat wider than the other, and between the two a greenish yellow part of the spectrum is distinctly seen.

As explained some pages back several reducing agents modify the oxyhæmoglobin in a very marked manner, which is readily shown by a change in the color of the diluted blood itself. The alteration as observed in the spectroscope is very striking and characteristic.

When treated with Stokes' solution dilute blood becomes purple in color and shows in the spectroscope one dark broad band filling three-fourths of the space between D and E, instead of two bands. If ammonium sulphide is used instead of the Stokes' solution the same broad band appears, and in addition a single narrow black band, the center of which is to the left of D.

Finally, diluted blood saturated with carbon monoxide shows two dark bands, differing in position, however, from those of oxyhæmoglobin. The space between them is narrower and they are both moved toward the blue of the spectrum. The fainter of these bands reaches nearly to the Fraunhofer line *b*, while the heavier one does not reach

D in the other direction. These appearances can be best noted by the student with solutions treated as in the following experiments:

Ex. 94. To a dilute solution of blood, about 1 part to 50 of water, add a few drops of strong ammonium sulphide solution and warm gently in a test-tube until the change of color noted above is reached. Now place the tube before the slit of the spectroscope and observe the bands referred to, especially the narrow one in the red.

Hydrogen sulphide gives practically the same result.

Ex. 95. Repeat the above experiment, using Stokes' solution instead of the sulphide. A single broad band appears now; if the liquid is shaken briskly the air acts on the reduced coloring matter with oxidizing effect, as shown by a division of the band, but only temporarily. On standing a short time the single broad band, not very sharply defined, returns.

Ex. 96. Into diluted blood, as before, pass a stream of common illuminating gas until the liquid is saturated, which requires but a few minutes. On placing the tube in front of the spectroscope the two dark bands described will be seen and farther from the yellow than is the case with oxyhæmoglobin.

These bands do not change in extent or position by agitation of the liquid with air, as follows with reduced hæmoglobin.

Methæmoglobin. This is derived from oxyhæmoglobin by action of certain oxidizing agents. It is said to be produced in the living body by excessive doses of potassium chlorate, and appears also, it is said, in several diseases.

The formation of this body can be illustrated by experiment:

Ex. 97. To diluted blood in the test-tube, (1 to 50), add some small crystals of potassium chlorate, and warm

106 *ELEMENTARY CHEMICAL PHYSIOLOGY.*

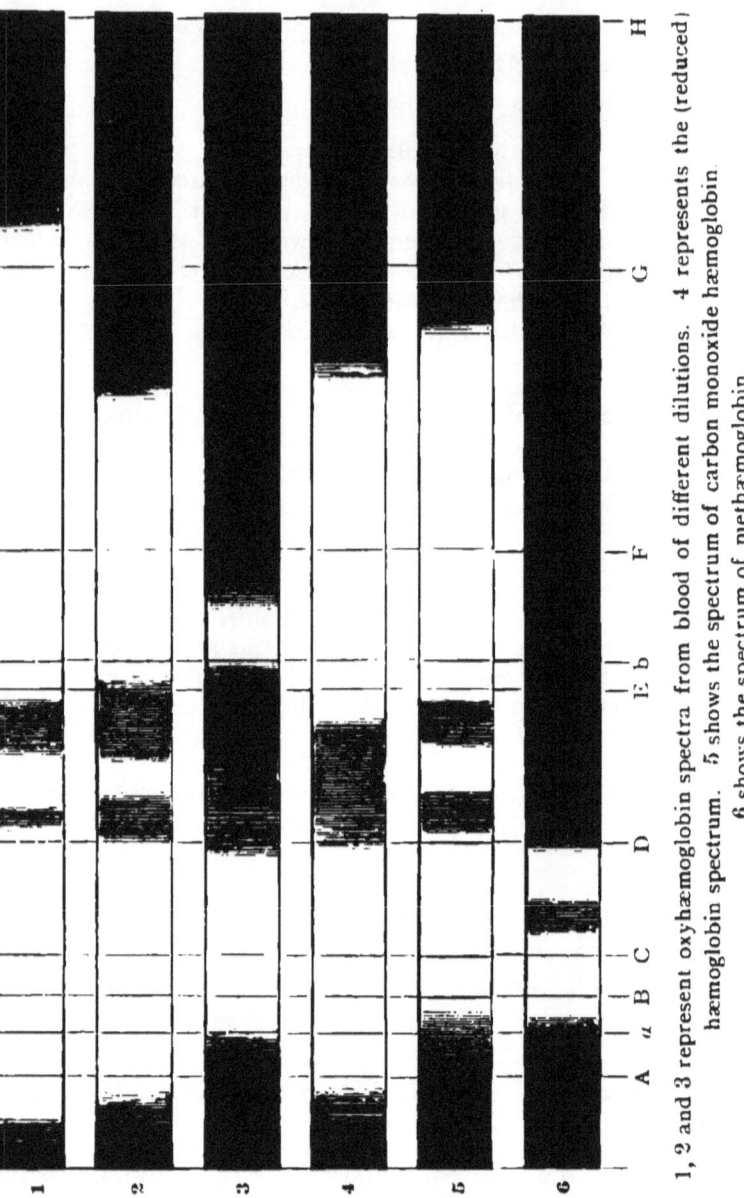

FIG. 32.

1, 2 and 3 represent oxyhæmoglobin spectra from blood of different dilutions. 4 represents the (reduced) hæmoglobin spectrum. 5 shows the spectrum of carbon monoxide hæmoglobin. 6 shows the spectrum of methæmoglobin.

very gently. To a like amount of blood in another tube add a few drops of a 1 per cent solution of potassium permanganate and warm. Both solutions should darken, and when examined by the spectroscope should show a number of characteristic bands, especially one in the red. Ammonium sulphide gives with this the spectrum of hæmoglobin as shown before.

The several spectra referred to are shown in the cut on the preceding page.

The clinical importance of variations in the amount of oxyhæmoglobin is so great that besides the approximate methods of measurement given above, other very elaborate and accurate ones have been devised, especially by *Vierordt*, which give results of scientific exactness.

The method of Vierordt is carried out by the aid of an instrument known as a spectrophotometer, illustrated in a following chapter.

The principle involved is this: Diluted blood, in a layer of definite thickness, absorbs a certain fraction of the light passing through it, and this fraction depends on the amount of hæmoglobin present, but is not the same for all parts of the spectrum. If the amount of absorption in a given spectral region is determined once for all with solutions of *known* strength, that is if the relation between concentration and absorption is found by direct experiment, it will be possible to find the hæmoglobin strength of an *unknown* solution by simply determining its absorption power for the same part of the spectrum.

Vierordt and others have shown how these tests may be made with great accuracy; and as the method is applied to many physiological investigations besides the investigation of blood it will be explained in some detail in a following chapter to which the student is referred.

The Number of Blood Corpuscles.

In the normal blood of man there are about 5,000,000 red corpuscles in the cubic millimeter. In the blood of men the number is somewhat greater than with women.

In several diseases the number may suffer a very great decrease, and as a means of diagnosis the determination of the condition of the blood in this respect becomes often of the highest importance. Usually, a change in the total amount of hæmoglobin may be taken as proportional to a change in the number of corpuscles, but this does not always hold true, and it is, therefore, necessary to make independent determinations.

A simple clinical method is carried out essentially in this manner. A small known volume of blood, accurately measured by a fine pipette, is diluted to a definite larger volume by addition of water. Then in a definite small fraction of this diluted mixture the corpuscles are counted under the microscope. With properly constructed measuring appliances it is possible to dilute 5 cubic millimeters of blood with 995 of water, or even 1 of blood with 999 of water. In order to effect an accurate count under the microscope it has been found most convenient to place a drop of the diluted blood in a shallow cell, having a depth of exactly one-fifth or one tenth millimeter. The bottom of this cell is ruled in squares a tenth or twentieth of a millimeter on each side. When the cell is filled with the diluted blood and covered with an ordinary cover glass the number of corpuscles between the rulings can be readily counted under the microscope. Counting instruments have been devised by *Gowers*, *Abbe*, and others. The *Gowers* instrument is shown in the annexed illustration.

In using this instrument 995 cubic millimeters of sodium sulphate solution (with a specific gravity of 1.015) are measured by means of the pipette A, and discharged into the mixing vessel D. 5 cubic millimeters of blood are drawn up by the pipette B, and mixed with the sulphate solution by means of a small glass stirrer. A drop of the mixture is put in the center of the glass slide, in the shallow cell the small squares in which are ruled in tenths of a milli-

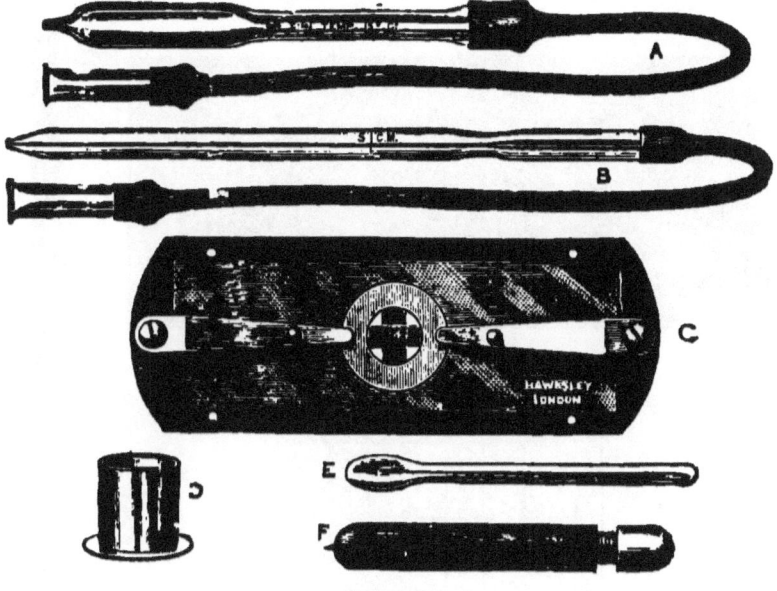

FIG. 33.

meter. The drop is covered with a cover glass which is pressed down. As the cell is one-fifth of a millimeter deep the volume of blood in the space bounded by the slide, cover and ruling is $\frac{1}{500}$ of a cubic millimeter. Therefore, 500 times the number of corpuscles counted in each square gives the number in one cubic millimeter of the diluted blood, and 200 times this product the number in the original blood. It is best to count the number in 15 or 20 squares and take the average; this multiplied by

100,000, gives the number per cubic millimeter in the undiluted blood. Normally the average number of corpuscles in each square is 50.

In the instrument made by *Reichert*, of Vienna, the *Thoma-Zeiss*, the cell is one-tenth of a millimeter deep,

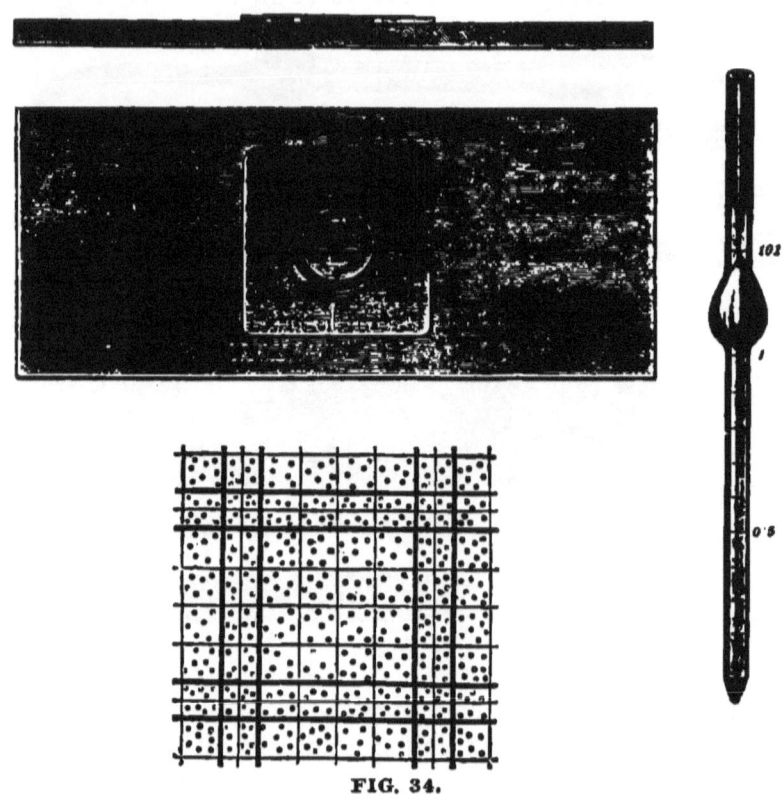

FIG. 34.

and the space bounded by the slide, cover and lines $\frac{1}{4000}$ of a cubic millimeter. The blood is drawn up in a mixing pipette which dilutes it 100 or 200 times. The reading with the dilute blood for each square must be multiplied by 40,000 or 80,000 to get the number of corpuscles in each cubic millimeter of the original. Fig. 34 shows the cell and the rulings and the measuring pipette.

Chapter VI.

BONE CONSTITUENTS.—SALIVA.—GASTRIC JUICE.—THE BILE.

A FEW simple tests may be readily carried out by the student to show the general composition of bones. Roughly speaking, bones consist of one part of organic matter to two of mineral matter. The organic substance is termed *ossein*. The mineral matters present show, in the mean, about the following composition :

    Calcium phosphate ....85.0 per cent.
    Magnesium phosphate............ 1.5   "
    Calcium carbonate................11.0   "
    Calcium fluoride and chloride..... 2.5   "

These mineral constituents are insoluble in water, but readily soluble in dilute hydrochloric acid, by the aid of which they may be separated from the ossein.

Ex. 98. Clean a long, slender bone (best, a rib), and immerse it in dilute hydrochloric acid of about ten per cent strength. Let it remain several days. At the end of this time remove the bone from the acid and observe that it has lost its rigidity, and has become very flexible. It may be even possible to tie it in a knot. Wash the elastic mass several times in fresh water to remove all the hydrochloric acid, and then boil it with a small amount of pure water. By heating it long enough the ossein becomes converted into gelatin which solidifies, on cooling, to a jelly.

By boiling the bone ossein under pressure the formation of the gelatin is very much hastened.

Ex. 99. Make a dilute aqueous solution of gelatin, and add to it an aqueous solution of tannic acid. This gives a white flocculent precipitate which is characteristic as it is produced even in extremely dilute solutions.

Commercially, gelatin occurs in shreds, or sheets, which are used for many purposes. It is very similar to albumin in chemical composition, although different in many important properties. It fails to give many of the reactions characteristic of albumins as a class, and does not appear to be able to take the place of albumins as a food.

In the laboratory gelatin is employed in the preparation of the well-known nutrient jellies used in the cultivation of bacteria.

In the next to the last experiment the mineral matters were left in the hydrochloric acid solution. Filter this, if it is not clear, and use it for tests.

Ex. 100. To a few cubic centimeters of the solution add some ammonium molybdate solution. In a short time a yellow precipitate appears, indicating presence of a phosphate.

Ex. 101. To a few cubic centimeters of the solution add solution of sodium acetate until a distinct odor of acetic acid persists. Then add some solution of ammonium oxalate which produces a white precipitate of calcium oxalate.

Ex. 102. To another portion of the hydrochloric acid solution add ammonia until a good alkaline reaction is obtained. A white precipitate of calcium and magnesium phosphates settles out. Filter and to the filtrate add some ammonium oxalate solution. A further precipitate appears. This is calcium oxalate and proves that the original bone contains calcium in excess of that necessary to combine with phosphoric acid. The calcium combined with carbonic, hydrofluoric and hydrochloric acids appears here.

Saliva.

In our experiments on starches an important property of saliva has been shown. A few other reactions remain to be given. Collect about twenty-five cubic centimeters of saliva (after thoroughly rinsing out the mouth) and filter it to obtain a clear solution, with which make the following tests.

Ex. 103. To a few Cc. of the clear saliva in a test-tube add several drops of a dilute solution of ferric chloride. This gives a more or less marked red color from the formation of ferric sulphocyanate. A very strong reaction must not be expected. Make a comparative test by adding a like amount of ferric chloride to dilute solutions of potassium sulphocyanate.

The addition of solution of mercuric chloride discharges the color.

A similar color is given by ferric salts and solutions of meconic acid, extracts of opium, for instance, and the reaction may therefore have medico-legal importance. Sulphocyanates are present in normal saliva and may, therefore, possibly be sometimes found in the liquids of the stomach in traces sufficient to give a test with the ferric salt when laudanum or other opium extract is looked for. Ferric sulphocyanate like ferric meconate is red, but danger of confounding the two may be avoided by noting that the color of the former is destroyed by mercuric chloride, while that of the latter is not affected.

Ex. 104. Test the reaction of saliva with *neutral* litmus paper. It will be found slightly alkaline. Now add two or three drops of dilute acetic acid and note that a stringy precipitate of mucin separates. Filter off this precipitate and test the filtrate for proteids by boiling with Millon's reagent or by the xanthoproteic reaction.

Mixed human saliva has a specific gravity of 1,002 to 1,006, and contains in the mean 99.5 per cent of water and 0.5 per cent of solids. The potassium sulphocyanate amounts to about 0.005 per cent.

Gastric Juice.

Chemical Experiments. Simple laboratory tests of the gastric juice cannot be readily made for want of a convenient source of the material. But this secretion has in the last three years become an object of accurate clinical investigation, since methods have been devised for collecting it without great annoyance to the patient. Investigations are usually instituted to determine the amount and character of the free acids and the peptic activity of the fluid. The latter test is carried out as already given. The acid tests will be explained here.

For investigation, gastric juice is collected at the time when the stomach is as free as possible from food, preferably before breakfast and by aid of the gastric sound and stomach pump or by a siphon tube. The sound for this purpose consists of an elastic rubber tube small and firm enough to be pushed through the œsophagus into the stomach. The lower end is closed and round but furnished with a number of very fine openings. By attaching the outer end of the tube to a stomach pump and exhausting, it may be partly filled with liquid. This is filtered and is then ready for the several tests.

Total Acidity. Measure accurately 10 Cc. of the filtered gastric juice into a beaker, add a few drops of phenolphthalein indicator and then from a burette add slowly $\frac{1}{20}$ normal sodium hydroxide solution until a pink color just persists on shaking. Each Cc. of the alkali solution used corresponds to 0.00182 Gm. of free hydrochloric acid, but it will not do to consider this as an estimation of hydrochloric acid as other acids are generally present in some amount, among them lactic.

One cubic centimeter of the $\frac{1}{20}$ normal sodium hydroxide solution neutralizes 0.0045 Gm. of lactic acid, or nearly two and one-half times the weight of the hydrochloric acid neutralized. The acidity must, therefore, be indicated in terms of alkali used and not expressed as acid found.

Hydrochloric acid is certainly the most abundant of the gastric acids and near the end of a digestive process it is said to be normally the only one present. But at the beginning of digestion lactic acid may be also present. Pathologically, lactic and other organic acids may be at times very much increased and their detection is a matter of no little clinical importance. A number of tests will here be given.

Free Hydrochloric Acid. That this acid is present in the uncombined condition has been frequently shown by making an accurate determination of all the bases and also of the total hydrochloric acid. The latter is in excess of the amount which could unite with the bases and must therefore be partly free. For the clinical detection of the acid the following methods are in favor.

"Emerald Green" Test. An aqueous solution of this substance, when added in small amount to weak hydrochloric acid, is turned yellowish green or yellowish brown. Organic acids do not give this test. The green used for this test must be the pure product of the Bayer laboratory, Elberfeld, having the empirical formula, $C_{27} H_{33} N_2 HSO_4$.

"Congo Red" Test. This substance in aqueous solution is turned blue by very dilute hydrochloric acid. Organic acids do not give the test.

The reaction is most conveniently carried out by means of test papers made by dipping filter paper in a solution of the coloring matter, and then drying.

"Methyl-Violet" Test. A dilute violet colored aqueous solution of this substance, when mixed with weak hydrochloric acid, turns blue. The reaction with gastric juice is faint, but when care is observed characteristic. Organic acids, even when present in quantity, do not give the test, which was first successfully used for the detection of traces of mineral acids in vinegar.

"oo Tropæolin" Test. When a dilute alcoholic or aqueous solution of this color is added to weak hydrochloric or other free acid the color changes from yellow to reddish violet. This is a sensitive test, especially for hydrochloric acid.

"Phloroglucin and Vanillin" Test. A reagent is made by dissolving 2 Gm. of phloroglucin and 1 Gm. of vanillin in 100 Cc. of alcohol. To make the test, mix 5 Cc. of this solution with an equal volume of the gastric filtrate, and concentrate in a glass or porcelain vessel on the water-bath. As the liquid becomes concentrated it turns red.

Tests for Lactic Acid. Prepare a dilute solution of phenol by dissolving 1 Gm. of the pure crystallized product in 75 Cc. of water. To this add 5 drops of a strong solution of ferric chloride, which produces a deep blue color. Five Cc. of this mixture suffices for a test. Add to it a few drops of the liquid containing lactic acid, and note the change from blue to yellow.

A weak, almost colorless solution of ferric chloride alone serves also as a test substance, as its color becomes much deeper by addition of a trace of lactic acid.

This reaction is not influenced by the presence of small amounts of hydrochloric acid, as can be readily shown by adding some to the liquid to be tested.

The Bile.

In the bile are found a number of characteristic pigments and acids which can be recognized without much difficulty and which serve, therefore, as indicators of the presence of the secretion.

Two acids, *glycocholic* and *taurocholic*, are found in human bile in combination as alkali salts; they are found also in bile of many animals. Investigations seem to show that in human bile the glycocholic acid is more abundant than the other.

Two important pigments are found in human and other bile normally, and these are known as *bilirubin* and *biliverdin*. Pathologically, it is likely that other pigments are present in small amount. Several modified forms of the two normal pigments have been described, which differ mainly in the amount of water of crystallization. The formula of bilirubin is $C_{16} H_{18} N_2 O_3$, while that of biliverdin is $C_{16} H_{18} N_2 O_4$.

In human bile bilirubin is probably in excess, while in ox bile the biliverdin seems to predominate. These pigments appear to be derived from the hæmoglobin of the blood, bilirubin being in fact identical in composition with hæmatoidin, which is an iron-free derivative of hæmoglobin.

Tests for the Bile Acids. Several color tests are known here, the so-called *Pettenkofer* test being perhaps the most characteristic. The student can use ox bile for this and other reactions given below.

Ex. 105. Add a little cane sugar or some strong syrup to bile in a test-tube. Then pour in an equal volume of strong sulphuric acid in such a manner as to mix the liquids as little as possible. The acid may be allowed to trickle down the side of the test-tube and collect beneath the lighter bile. At the junction of the two liquids a dark purple band appears. On shaking the tube the liquids mix and become colored throughout. A modified form of the test is sometimes carried out in this manner: Mix a little syrup with bile and shake the tube until a layer of froth forms. Pour in a few drops of strong sulphuric acid. As it passes through the froth it imparts a purple color to it.

This reaction depends on the production of furfurol ($C_4 H_3 O C H O$) by the destruction of the sugar when the sulphuric acid is added. Furfurol in turn combines with cholalic acid, formed by the action of the sulphuric acid on the bile acids, giving the color. Several substances give a very similar color, and the test, therefore, must be employed with caution. With very dilute solutions the reaction does not appear and this is the case when the test is directly applied to urine. In order to avoid all uncertainty, either from dilution of the solution to be examined, or from possible presence of interfering substances it is necessary to apply a method for the separation of the bile acids, and on these perform finally the Pettenkofer test. Under "Urine Analysis," later, such a method will be explained in detail.

Another modification of the test consists in diluting the colored solution and observing its absorption bands in the spectroscope. The number and position of the bands given by cholalic acid are said to be quite characteristic.

Ex. 106. The Pigment Tests. These, as usually carried out, depend on the oxidation of the pigments by means of nitric acid. *Gmelin's* test is performed in this manner. To a few Cc. of strong nitric acid in a test-tube add a little bile, without mixing much. At the junction of the two liquids a series of colored rings appear, green, blue, violet, red and yellow below. Try the test, also, by placing some bile in a flat-bottomed porcelain dish. When a drop of strong nitric acid is put in the middle of the bile a play of colors is observed as in the test-tube. The oxidation is greatest, with yellow color, near the acid, and least, with green color, near the bile.

This reaction is exceedingly delicate and is applied chiefly to the detection of bile in urine.

Other constituents of the bile are shown by the following experiment:

Ex. 107. To 5 Cc. of bile add an equal volume of water and some alcohol. This produces a precipitate of mucin. Filter this off and divide the filtrate into two portions; to one add some hydrochloric acid which causes precipitation of glycocholic acid, to the other portion add solution of lead acetate, which throws down lead glycocholate. Remove this by filtration, and to the filtrate, add solution of *basic* lead acetate, which gives a further precipitation of lead taurocholate.

Action on Fats. In Chapter III. it was shown that emulsions of fats are produced in various ways, especially by action of the pancreatic juice and by the bile. This latter reaction can now be illustrated by experiment.

Ex. 108. In a slightly warmed mortar pour about 5 Cc. of bile, and add to it 1 Cc. of cottonseed oil. Rub the two thoroughly together for several minutes, and then add another small portion of the fatty oil. An emulsion forms slowly, and becomes more persistent as the working with the pestle is prolonged. The amount of oil which can be brought into the form of a stable emulsion with the 5 Cc. of bile depends largely on the character of the oil. The presence of a small amount of free fatty acid in the cottonseed oil aids materially in producing the emulsion.

The free fatty acids have the power of decomposing the bile salts with liberation of their acids. The soaps formed assist in increasing and holding the emulsion.

It is also worthy of note that animal membranes moistened with bile permit the passage of fatty oils, while if they are moistened with water only the oil cannot pass through. This behavior is of the highest importance in aiding the absorption of fatty substances from the intestine in the process of digestion.

Chapter VII.

MILK.—BEEF EXTRACTS.—FLOUR AND MEAL.

FROM its very great importance milk has been made the subject of almost countless investigations from nearly every standpoint. Most of the literature is naturally concerned with cow's milk as a commercial article having market value. Besides this, however, we have a valuable scientific literature of milk which discusses its secretion as a physiological process and variations in its composition depending on age, race, nutrition, etc., of the animal furnishing it.

For many reasons our knowledge of human milk is far less complete than is our knowledge of the milk of several animals. The difficulty of collecting normal human milk is naturally very great, as any stimulus applied to cause its flow for collection must necessarily be an abnormal one. The wide disagreement between many of the published analyses of human milk may probably be in part accounted for from this fact. Other reasons will be pointed out below.

The average composition of cow's milk as determined by thousands of analyses is given in the following figures:

Water..........................87.4 per cent
Fat............................. 3.5   "
Sugar.......................... 4.5   "
Albumins....................... 3.9   "
Salts.......................... .7   "

The solids amount to 12.6 per cent.

In considering milk furnished by individual cows in nearly 400 instances the following variations depending on age, race, season, food, etc., were noticed by a well-known German authority :

The solids between 8.50 and 16.03 per cent, the fat between 2.04 and 6.17, the sugar between 2.00 and 6.10, the albuminoids between 1.98 and 6.61, and the salts between 0.34 and 0.98 per cent.

The following are some analyses which have been given of human milk. (*Landois and Stirling.*)

Water	87.24 to	90.58	per cent.
Fat	2.67 "	4.30	"
Sugar	3.15 "	6.09	"
Albumins	2.91 "	3.92	"
Salts	.14 "	.28	"

Recent analyses by *Palm*, made by methods which seem to be reliable and free from errors of older methods, give the following results as the mean of twenty complete analyses of nurse's milk :

Water	87.81	per cent.
Fat	4.06	"
Sugar	5.26	"
Albumins	2.36	"
Salts	.51	"

Human milk is poorer than cow's milk in albumins, but richer in sugar.

Some simple experiments may readily be made to show the presence of the three important constituents in milk.

The Test for Fat.

Ex. 109. Pour about 20 Cc. of milk in a porcelain dish, add an equal volume of clean, dry quartz sand and evaporate, with frequent stirring, about an hour on the water-bath. Then loosen the dry mass as well as possible by means of a spatula, or glass rod, and pour over it 25 Cc. of light benzine. Stir up well and cover with a sheet of paper and allow to stand 15 minutes. Then pour the liquid through a small, dry filter into a small, dry beaker, and place this in hot water to volatilize the benzine.

A residue of fat will be left. *Do not attempt to evaporate the benzine over a flame, or on a water-bath under which a lamp is burning.* Heat the water, then extinguish the flame and immerse the vessel containing the benzine in the hot water.

In the analysis of milk the determination of the amount of fat is the most important operation, and many quick and accurate methods have been devised by which this may be accomplished. The processes by drying and extraction by ether, carbon bisulphide or benzine, as illustrated above, can be made very exact, but then they consume a great deal of time. It has been found possible to mix the milk with certain reagents which cause the fat to separate in a pure layer, and if the separation takes place in a narrow, graduated tube, the volume of this fat layer may be read off accurately. The specific gravity of the fat being known, the weight of a given observed volume is obtained, and from this the percentage amount. Several forms of apparatus are now in use by means of which this may be easily done, so that it is possible to make many tests in a day, or in an hour even.

The fat residue is a mixture of the glycerides of oleic, palmitic, stearic, butyric and other acids. The above method can be made quantitative by weighing the milk, drying carefully, extracting completely and drying and weighing the fatty residue.

The Test for Sugar.

Ex. 110. Measure out about 10 Cc. of milk, and dilute it with water to make 200 Cc. Add to this 5 Cc. of a copper sulphate solution such as is used in making the Fehling solution, (69.3 Gm. per liter) and then enough potassium or sodium hydroxide solution to produce a voluminous precipitate containing copper with all the proteids and fat. For this purpose about 3.5 Cc. of a 1 per cent sodium hydroxide solution will be required. Allow the precipitate to subside, pour or filter off some of the supernatant liquid, and boil it with Fehling solution. The characteristic red precipitate forms, showing presence of sugar.

The sugar appearing in the above test is known as milk sugar or lactose. Like cane sugar it suffers inversion when heated with weak acids, the product formed in this way being known as galactose.

The principle illustrated by the above experiment is readily made the basis of a quantitative process of value.

It has usually been assumed that the precipitation of proteids and fat is complete by this reaction, but there is good reason for believing that certain modified albumins are not thrown down in this way, but are left with the filtrate to slightly impair the accuracy of the sugar determination. The precipitate has frequently been used for the determination of proteids after drying and dissolving out the fats, but the results are probably a little low, for the reason just mentioned, especially in the case of human milk.

Proteid Test.

The presence of a proteid in milk can readily be shown as follows:

Ex. III. Mix equal volumes of milk and Millon's reagent in a test-tube, and boil. The bulky red precipitate which forms, proves the presence of the body in question.

The total amount of proteid present in milk can be most accurately determined by finding the nitrogen by the methods of organic analysis and multiplying this by 6.25, on the supposition that proteids in the mean contain 16 per cent of nitrogen. That milk contains casein as its chief nitrogenous constituent has already been shown in the chapter on proteids. It was also shown that a simple albumin is likewise present, which can be coagulated after separation of the casein. Peptone appears to be present in small amount as a rule and this escapes precipitation by the usual methods.

Action of Rennet on Milk.

The mucous membrane of the stomachs of most animals, and especially that of the young calf, contains an enzyme known as the "milk curdling ferment," the "rennet ferment" or rennin.

A crude extract of the stomach mucous membrane from the calf is commonly called "rennet" and has long been in use for the curdling of milk in the production of cheese. This curdling consists essentially in the coagulation or precipitation of the casein, which it will be recalled, is not readily thrown down by the usual methods.

An active rennet can be readily obtained by digesting the fourth stomach of the calf with glycerol or brine. If a brine extract is precipitated by alcohol in excess a white powder separates, which, when collected and dried, has very active properties. Several powders of this description are now in the market. Let the student try the following experiment with such a product:

Ex. 112. Warm some fresh milk to a temperature of 38° to 40° C. in a test-tube or small beaker, then add about half a gram of commercial "rennin," and after stirring it in well keep for fifteen minutes at a temperature not above 40°. Then as the milk cools it assumes the consistence of a firm jelly. It is essential in this experiment that the temperature be kept within the proper limits, as the enzyme is not active at low temperature and it is, like others, destroyed by high temperature.

The spontaneous coagulation of milk is due to the action of lactic acid produced by the conversion of milk sugar under the influence of a true ferment which enters the milk from the air usually. Boiling destroys this ferment. Milk when heated to 100° for an hour in bottles loosely stoppered with cotton plugs may be kept sweet almost indefinitely provided the plugs of cotton are not removed.

Milk which has been boiled and allowed to cool does not coagulate readily with rennet. A perfectly satisfactory explanation of this fact has not been given.

The spontaneous coagulation of boiled milk is usually very slow, as the necessary lactic acid ferment cannot be supplied by all atmospheres.

The Action of Pancreatic Extract on Milk.

The behavior of milk with extract of pancreas is somewhat complicated because of the complex nature of milk itself. The three important constituents of milk, the sugar, the fat, and the proteid bodies all suffer some change under the influence of the several pancreatic enzymes.

The most interesting of these changes, however, is that produced in the proteids, and is commonly called peptonization.

At the present time the digestion, or peptonization of milk, is a very common practice in the preparation of food for the sick room, and can be illustrated by the following experiment :

Ex. 113. Dilute about 10 Cc. of milk with an equal volume of water, and add half a gram of sodium bicarbonate. Next add a few drops of a liquid extract of pancreas, or a very small amount (10 to 20 Mg.) of one of the concentrated "pancreatin" powders on the market. Shake the mixture and keep it at a temperature of 40° on the water-bath half an hour. At the end of this time filter and apply the peptone test—potassium hydroxide and dilute copper sulphate—and observe the pink color.

The following formula will serve for the practical digestion of milk in quantity:

Dissolve 1 Gm. of sodium bicarbonate in 100 Cc. of water, and add from a third to a half gram of pancreatin powder. Warm the mixture slightly, not above 40°, and add 500 Cc. of milk, warmed to about 40°. Keep the mixture at this temperature about half an hour. As the action goes on the color changes from white to grayish yellow, and a bitter taste appears which becomes stronger the longer the digestion is continued. As this bitter taste is unpleasant, it is always necessary to stop the reaction before it is fully completed which can be done either by boiling the milk, or by cooling it quickly by placing it on ice. If it is to be used immediately it is not necessary to boil or cool.

The pancreatin, or pancreatic extract, used for this purpose, must be from beef, not from the hog pancreas. An extract from the latter source is very active in the conversion of starch into sugar, but is deficient in proteid converting power. Some manufacturers prepare products from both sources.

Extract of Meat.

By the term *extract of meat* several different products may be meant. When lean meat is superficially broiled, minced fine and squeezed out in a meat press, a juice is obtained which holds, besides meat salts and meat bases, a certain amount of soluble albumin. Such a product has considerable direct nutritive value. If, however, this juice is thoroughly boiled and filtered practically all the albumin is coagulated and lost, and a product so made would have little nutritive value.

It was at one time supposed that a concentrated "extract" of meat could be secured by boiling meat until completely disintegrated, filtering and evaporating the filtrate to a paste, and that in this manner the most valuable part, or *essence* of the beef was secured. This view was afterward shown to be fallacious, and the opinion that the product so prepared contained only the meat salts and organic bases, is therefore merely a condiment and stimulant, and not a true food in any sense, took its place.

The error here seems to be as great as in the other case. It has lately been found possible to effect a separation of the derived proteid products known as albumóses and peptones, described in a former chapter, and a study of meat extract shows that these bodies are often present in considerable proportion. The amount present depends on the time of boiling, and on some other factors.

It seems that by prolonged contact with hot water a partial digestion of the meat takes place so that from the coagulated albumin, albumose, and finally peptone is formed. In the boiling operation any fat present would separate so that it could be skimmed off. Among the meat bases present there may be mentioned *carnin, kreatin* and *sarkin.* Some gelatin is always present too.

The production of extract of meat was first undertaken in order to utilize the flesh of cattle killed on the plains of South America for the hides. It is now made elsewhere in great quantities. Two samples recently analyzed by the author showed the following constituents:

Solid matter	80.4 per cent.	80.7 per cent.
Water and volatile	19.6 "	19.3 "
Ash	20.2 "	24.0 "
Soluble albumin	trace. "	trace. "
Insoluble albumin	0.0 "	0.0 "
Albumose	12.7 "	4.3 "
Peptone	8.4 "	8.9 "
Other N bodies	39.1 "	43.5 "

From its composition it is apparent that alone it cannot be used as food, but as a condiment and addition to other food. From this standpoint it has great value.

Several of the organic basic bodies present are active stimulants, and the potassium phosphate and other salts of the same metal are not without marked physiological properties.

A few simple experiments will show important characteristics of the commercial extracts as everywhere found on the market.

Ex. 114. Heat a little of the solid extract on a bit of porcelain until it is reduced to a char. Extract this with dilute nitric acid, filter and divide the filtrate into two portions. In one, test for phosphates by the addition of ammonium molybdate, and in the other, for potassium salts by the flame test. Both tests should show good reactions.

Ex. 115. Add 200 Cc. of water to 10 Gm. of commercial extract, warm gently and observe that a nearly clear solution is obtained, showing absence of fat, coagulated albumin, etc. To the solution add very carefully a solution of basic acetate of lead as long as a precipitate forms, but not much in excess. This can be determined by waiting after each addition until the precipitate settles enough to leave a moderately clear supernatant liquid. A few drops of the lead solution added to this will show whether more is needed or not. When precipitation is complete the light colored mass of lead salts, organic and inorganic, is filtered off, and through the filtrate enough hydrogen sulphide is passed to throw down the excess of lead contained there. The black sulphide is removed by filtration, and the filtrate shaken thoroughly to remove as much as possible of the gas. It is then evaporated at *a low temperature* on the water-bath to a volume of about 8 Cc., and allowed to stand then two or three days in a cool place. *Kreatin* separates as a crystalline mass. Pour the liquid and crystals on a filter, and wash with strong alcohol, in which kreatin is but slightly soluble.

Ex. 116. Dissolve the kreatin crystals of the last experiment in a small amount of pure hydrochloric acid, and evaporate the solution to dryness on the water-bath. By this action kreatin is converted into *kreatinin*. Dissolve this residue in a small volume of water, and divide the solution into two parts. To one add a solution of zinc chloride, which produces a white crystalline precipitate, the character of which is best seen under the microscope. With the other try *Weyl's* reaction. Add a few drops of dilute solution of sodium nitroprusside, and then, a drop at a time, dilute sodium hydroxide solution. This gives a ruby red color which soon fades to yellow. Add now enough acetic acid to change the reaction, and warm. The color becomes green, and finally bluish.

Kreatinin is interesting as occurring normally in urine to the amount of about 1 Gm. daily. It is probably derived there by dehydration from the kreatin found normally in muscle, blood and brain, as they differ in composition simply by one molecule of water.

The average composition of the commercial extract, corresponding to the analyses given, is about:

```
Water..............................20 parts.
Salts............................. ...20   "
Organic substances...................60   "
```

It is made on the large scale by boiling lean meat until everything soluble has passed into solution. The liquid is filtered and concentrated to the above composition, after removing fat by skimming, best in vacuum pans.

The water in which meat is boiled in the canning establishments is now generally used also in the production of extract, as it becomes concentrated after a time. When extract is made as the principal product the muscular residue remaining after long boiling has little value except as a cattle food, for which purpose it is sometimes employed.

Fresh beef is sometimes digested with pepsin and hydrochloric acid, or with pancreatic extract and soda and the product neutralized. Such products are sold as "fluid beef" or "peptonized meat," and under other names. Still other commercial articles seem to be prepared by simply heating meat with water and dilute hydrochloric acid under pressure. A preparation of this description contains proteids in a finely divided condition, but cannot be called digested. There are great differences in the value of these products as articles of food. In Chapter IX. further details concerning the examination of these products will be given.

Flour and Meal.

The average composition of wheat flour may be shown by the following analyses:

	Fine flour.	Coarse flour.
Water	13.34	12.65
Proteids	10.08	11.82
Fat	0.94	1.36
Sugar and gum	5.41	5.95
Starch	69.44	66.28
Fiber	0.31	0.98
Ash	0.48	0.96
	100.00	100.00

Corn meal contains usually less water and more fat than wheat flour.

Ex. 117. Boil a small amount of wheat flour with Millon's reagent. The red color produced shows presence of proteids.

Ex. 118. Moisten about 25 Gm. of flour with water and work it into a dough. Then hold this under a fine, slow stream of water and by kneading between the fingers slowly work out a portion of the mass as a thin milky liquid. This is largely starch. After some time an elastic residue is left insoluble in water. This is *gluten*, and is the chief nitrogenous element of the flour.

Gluten corresponds to fibrin of the animal proteids and is accompanied by a vegetable albumin and vegetable casein, besides other related products in small amounts.

Ex. 119. To about 5 Gm. of flour add 10 Cc. of water, shake thoroughly and allow to stand until a nearly clear liquid appears above a white sediment. Filter the liquid and test for sugar by the Fehling solution.
Boil some of the residue with water and add iodine solution as a test for starch.

Ex. 120. To about 5 Gm. of fine corn meal in a test-tube add 10 Cc. of ether. Close the tube with the thumb and shake thoroughly. Then cork and allow to stand half an hour. Shake again and pour the mixture on a small filter, collect the ethereal filtrate in a shallow dish and evaporate it by immersion in warm water. A small amount of fat will remain.

Rice flour is characterized by containing a large amount of starch with only a small amount of fat, and much less albuminous matter than is found in corn or wheat. Peas and beans are rich in albuminoids with lower carbohydrates than corn or wheat.

ACTION OF YEAST ON FLOUR.

The following experiment is intended to illustrate the work done by yeast in leavening dough:

Ex. 121. Crumble two or three grams of compressed yeast into 15 Cc. of lukewarm water and shake or stir the mixture until the yeast is uniformly distributed. Then stir in enough flour to make a thick cream and allow to stand over night at room temperature. In this time fermentation of the small amount of sugar in the flour begins and the "sponge" swells up by the escape of bubbles of gas. At this stage mix in uniformly and thoroughly enough flour to make a stiff dough, using for the purpose perhaps 25 Gm. Put the dough in an evaporating dish, keep it for an hour or more at a temperature of 30° to 35° C. and observe that it increases very greatly in size, from the continued action of the yeast in liberating bubbles of carbon dioxide. If a *good* hot air oven is at hand the experiment is completed by baking the leavened mass.

Fresh brewer's yeast, if obtainable, is preferable to the compressed yeast for this experiment as its action is quicker.

Yeast is sold also in the form of perfectly dry cakes which keep almost indefinitely and give most excellent results.

Chapter VIII.

WATER AND AIR.

THE sanitary examination of water is a matter of some difficulty and can be carried out properly only by skilled chemists. But certain tests are so commonly used and certain terms so frequently employed that it will not be out of place here to illustrate by a few experiments the nature of the tests and meaning of the terms.

Absolutely pure water is nowhere known as a natural substance, but can be prepared only by elaborate methods of distillation.

Natural waters secured from springs, wells, rivers or lakes hold in solution or suspension a variety of substances taken from the air or from the soil through which they have passed. Many of these substances are harmless. In fact, they may be even beneficial in a drinking water, and with them we are not concerned here. Certain other substances, which in themselves are harmless, are generally looked upon with suspicion in drinking waters because they usually enter water accompanied by other substances of really dangerous character, or because they are products of decomposition of possibly dangerous substances. Thus, common salt is found in small amount in nearly all waters, but in the water of ordinary wells and rivers does not exist in more than traces because soils are not strongly impregnated with salt. If in a shallow well water we find more than a trace of salt we are led to look for the source from which it has come.

The urine and solid excreta of common house sewage are the worst offenders in the contamination of well water where they are allowed to soak into the soil through defective drains or improperly constructed vaults.

Most of the organic products in the sewage may be speedily oxidized by the soil. Others may pass on and enter the water, and with them the indestructible salt as an indicator of past contamination, as a reminder of the possible presence of something else for which we have less characteristic tests.

The test for salt is a very simple one, and we make it in our search for possible past contamination.

We test for ammonia, for nitrates, and for nitrites, not because the minute quantities of these bodies found in waters are harmful, but because they are usually produced by the decomposition of nitrogenous organic matters and may be accompanied by germs of disease from the same source, from fæcal matter, for instance.

Ex. 122. The Test for Chlorides. A test is often made in this way. Measure out 200 Cc. of the water, add to it a few drops of a solution of pure neutral potassium chromate, and then from a burette run in, with constant stirring, solution of tenth normal silver nitrate until a faint reddish precipitate of silver chromate appears. Each cubic centimeter of the silver solution precipitates 3.54 Mg. of chlorine from common salt or other chloride, and when the last trace of chlorine is combined, the silver begins to precipitate the chromate with production of a red color. The chromate acts here as an *indicator*, as it shows just when the chlorine is all combined by beginning to precipitate itself.

In making this test it is well to take two similar beakers, place them side by side on white paper, pour equal amounts of water in each, add to each the same number of drops of the indicator, and then with one make the actual test by adding the silver solution. Note the amount used to give a light shade and then discharge it by adding a drop of salt solution. Now, with this opalescent or turbid liquid for comparison add silver nitrate to the second beaker until the light yellowish red shade just appears. This reading is usually somewhat more accurate than the first.

The amount of chlorine in most uncontaminated waters is less than 20 milligrams in a liter. The preparation of the standard silver nitrate solution, and of other solutions to follow, is given in the appendix.

Ex. 123. The Test for Ammonia. Solutions of ammonia or ammonium salts possess the peculiar property of giving a yellowish brown color with what is known as Nessler's reagent (a solution of mercuric potassium iodide, made strongly alkaline with sodium or potassium hydroxide). With more than traces of ammonia a precipitate is formed.

To make the test measure out 50 Cc. of the water in a large test-tube, or tall narrow beaker, and add to it 2 Cc. of the Nessler solution. By placing the beaker on a sheet of white paper and looking down through it, the depth of color can be observed. A few parts of ammonia in one hundred millions can be readily seen and measured.

Chemists usually make this test by measuring out 500 Cc. of the water, which is made alkaline by the addition of a few drops of strong, *pure* solution of sodium carbonate, and then distilled from a large, clean glass retort with Liebig's condenser attached. The distillate is collected in portions of 50 Cc. each in a number of thin cylinders of colorless glass, and to each is added 2 Cc. of the Nessler reagent. Four portions are usually enough, as the ammonia distills over easily and soon. The colors are duplicated by adding to pure distilled water in similar tubes small amounts of standard ammonia solution and then the

Nessler solution until like shades are obtained. In this manner it is possible to make a quantitative test.

In waters containing relatively large quantities of ammonia much less than 500 Cc. must be taken. Whatever the volume is which may be decided on by an approximate preliminary experiment, it should be diluted to 500 Cc. with pure ammonia-free water and then distilled.

A large amount of ammonia is generally an indication of contamination, but not always. Deep well waters often contain relatively great quantities of ammonia, while at the same time they may be organically pure.

Chemists apply the term *free ammonia* to that distilled as just explained. If to the residue in the retort after the distillation of 200 or 250 Cc. a strong oxidizing mixture of potassium or sodium hydroxide and potassium permanganate be added and heat again applied, a new portion of ammonia may be liberated and collected with the condensed steam as before. To this the term *albuminoid ammonia* is applied, because albuminous and other nitrogenous matters are broken up by this treatment with liberation of ammonia. The ammonia collected in the distillate during this operation did not therefore exist already formed in the water in the "free" or saline condition, but potentially. That is, its elements were present in complex nitrogenous bodies which, possibly, by putrefactive or other process of disintegration would yield it. The presence of these complex nitrogenous compounds which yield ammonia is a suggestion of the possible presence of worse matters, hence the value of the test.

Waters furnishing more than 1.5 to 2 parts of albuminoid ammonia in ten millions are usually condemned or looked upon as suspicious.

The Oxidation Tests. Pure waters absorb free oxygen from the atmosphere but have no tendency to decompose compounds to secure it. On the other hand waters containing organic matters or certain inorganic contaminations have the power of decomposing oxygen salts to secure the oxygen they desire, and the amount of oxygen so taken up becomes a measure of the impurity of the water. Potassium permanganate is a salt, which, under certain conditions, gives up its oxygen to waters containing organic bodies in solution and is frequently employed in water analysis for this purpose. An experiment will show one way in which it is used.

Ex. 124. Measure out about 100 Cc. of pure, carefully distilled water, pour it into a clean beaker in which water has just been boiled and add 5 Cc. of pure dilute sulphuric acid (1 to 3). Place the beaker on wire gauze and heat to boiling. Now add 5 drops of a dilute permanganate solution (300 milligrams to the liter) from a burette or dropping tube and boil five minutes. The pink color persists.

Repeat the experiment using 100 Cc. of common hydrant water to which a trace of egg albumin or urea has been added, and after running in the permanganate boil again. The color fades out and more may be added. Finally, after sufficient has been added the pink color remains. The number of drops or cubic centimeters used is a measure of the contamination of the water, although often, as in this experiment, a very rough one.

The above experiment is intended to show only the principle involved in the test. As used by chemists practically, it has been modified in several ways in order to make it convenient and reliable.

Because we are seldom in condition to tell the exact nature of the organic bodies present in a contaminated water, and are therefore unacquainted with their molecular weights or powers of combination we cannot express the results obtained by our tests in terms of organic matter present, but only as *oxygen consumed*.

In presence of organic matter (to absorb oxygen) and sulphuric acid, permanganate is decomposed according to the following equation:

$$3\,H_2SO_4 + K_2Mn_2O_8 = K_2SO_4 + 2\,MnSO_4 + 3\,H_2O + 5\,O.$$

That is, 315 parts of permanganate liberate 80 parts of oxygen. In a solution containing 315 milligrams of permanganate to the liter, each cubic centimeter will liberate 0.08 Mg. of oxygen.

The test is usually made by adding at once what a preliminary experiment shows to be an excess of permanganate, boiling five minutes and then determining the actual excess of permanganate by running into the hot liquid a solution of oxalic acid of such strength that one cubic centimeter will exactly reduce the same volume of permanganate.

The Tests for Nitrites and Nitrates. Nitrogenous matters undergoing oxidation in water and soil usually give rise, in time, to nitrites and finally to nitrates. These compounds are therefore looked for in a water as evidence of past contamination. In most instances nitrites, as a less advanced stage of oxidation than nitrates, suggest comparatively recent contamination. The tests are especially interesting in the examination of well and spring waters.

Chemists are acquainted with a number of methods for the detection of traces of nitrogen in the form of nitrites and nitrates, but at the present time certain color reactions are, because of their simplicity, mainly in favor. These are illustrated by the following tests.

A reagent for nitrites is prepared by dissolving 0.5 Gm. of sulphanilic acid in 150 Cc. of acetic acid of 25 per cent strength, and mixing this with a solution of 0.1 Gm. of pure naphthylamine in 200 Cc. of dilute acetic acid. This mixture keeps very well for a time in the dark.

Ex. 125. To about 50 Cc. of water in a clear beaker add 2 Cc. of the above solution. If the water is quite free from nitrites the reagent imparts no color to it. One hundredth of a milligram of nitrogen as nitrite in the water gives a faint pink color at the end of five minutes; with larger quantities the color may become deep rose red.

A fairly accurate quantitative measurement may be made by comparing the colors obtained with those produced in pure distilled water containing known amounts of added potassium or sodium nitrite when treated with the same reagent. The tests are best made in clear glass cylinders with flat bottoms. All waters allowed to stand in the air a short time show the reaction from absorbed nitrous acid.

Nitrates are sometimes detected in water by converting them into ammonia by some reducing agent, e. g., by addition of aluminum wire and pure sodium hydroxide, after which the ammonia is identified and measured by the Nessler test. Among other good methods is one illustrated in principle by the following experiment. Prepare first a solution of phenolsulphonic acid by adding to 50 Cc. of *pure*, *strong* sulphuric acid, 4 Cc. of water and 8 Gm. of pure phenol.

Ex. 126. Evaporate 50 or 100 Cc. of water to dryness in a beaker, add 1 Cc. of the reagent and mix it with the water residue. Then add 1 Cc. of pure distilled water and a few drops (3 or 4) of pure strong sulphuric acid. Warm the beaker by placing it for a few minutes in hot water, then dilute the contents to 25 or 30 Cc., add ammonia to give a good alkaline reaction (as shown by the smell), and finally water enough to bring the volume of liquid up to 100 Cc. The presence of nitrate is shown by a more or less deep yellow color.

In this experiment picric acid is formed at first if a nitrate is present, and the ammonia makes ammonium picrate which has a strong yellow color.

When used as a quantitative test the color is duplicated by treating water containing known amounts of nitrate in the same manner until the same shade is reached.

The simple exercises outlined above, while not intended as directions for the carrying out of exact scientific processes, are probably sufficient to give the student a general idea of the principal chemical tests employed at the present time in sanitary water examinations.

While physicians are seldom called upon to make chemical examinations their opinions are frequently asked in explanation of chemical reports and the work of the last few pages will impart some degree of familiarity with the usual terms employed.

Tests of Expired Air.

Experiment has shown that expired air differs essentially from that taken into the lungs by containing much less oxygen and a greatly increased amount of carbon dioxide. The average volume of the latter gas expired is over one hundred times as great as the volume taken in with the atmosphere.

Expired air contains, also, traces of ammonia and organic gases and vapors, which sometimes have a very disagreeable odor.

Several interesting tests may be made by blowing air from the lungs through distilled water and testing the latter. In order to prevent the passage of saliva into the water, which would vitiate some of the tests, the air is made to pass through a small empty bottle or test-tube first, with tubes leading in and out as in an ordinary wash bottle, the tube from the mouth being the longer, however.

Ex. 127. Pour some pure water into two small flasks or beakers. To one add some clear lime water and to the other five drops of tenth normal sodium hydroxide solution and a few drops of phenol phthalein indicator which produces a red color. Blow air into both flasks. In the one containing the lime water a white precipitate appears from the formation of calcium carbonate. In the other the red color, characteristic of an alkaline reaction, soon disappears as the alkali added becomes saturated by the carbonic acid. On the second reaction is based a very simple and accurate method of measuring quantitatively the carbon dioxide in crowded rooms.

Ex. 128. Force the expired air into a small flask containing fifty Cc. of carefully distilled water, to which has been added 1 Cc. of potassium permanganate solution, such as was used in the water test (oxygen consumed), and a small amount of pure sulphuric acid (about 1 Cc. of acid, 1 to 3.). Heat the flask nearly to boiling, and observe that as the air is forced through, the pink permanganate color fades, and gradually disappears, owing to the reducing power of the organic vapors of the breath.

Ex. 129. Into 50 Cc. of fresh distilled water in a small, clean flask blow air five or ten minutes. Then add about 2 Cc. of Nessler solution to the water and notice the yellow color formed, indicative of the presence of ammonia. The amount of nitrogen given off from the lungs in the form of ammonia is minute, but can usually be recognized readily by the above test.

Before beginning any of the above experiments the mouth should be *thoroughly* rinsed out with water in order to remove any organic matter left from the food or from other source.

Chapter IX.

SPECIAL PROBLEMS.

IN this chapter a number of special exercises will be given of a more advanced character than are those of the preceding pages. They are intended for the use of students who wish to prepare themselves for independent investigation in chemical physiology, but still are of such a nature that they properly belong in a course of undergraduate study, as given in the best of our American schools.

Problem 1. The Preparation of Urea.

The preparation of urea is from several standpoints an exercise of interest and practical importance. Representing, as it does, the final stage in nitrogenous metabolism in the human body, a study of the methods by which urea can be produced and of the compounds into which it can be transformed must naturally lead to some knowledge of the reactions taking place in the body itself. Chemically speaking, a long distance separates the nitrogen of the food from the nitrogen of the urine, and the course from one to the other is being explored from both directions. In the laboratory urea may be prepared from urine itself, or by several synthetic reactions.

a. From Urine. Evaporate two to three liters of urine, on the water-bath, to a volume of about 100 to 150 Cc. Cool this residue to 0° by surrounding it with ice and salt. Then add 300 Cc. of pure nitric acid diluted to a strength of 50 per cent, and cooled likewise to 0°. The acid should be added slowly with stirring, after which the mixture is allowed to stand several hours, or over night, at a low temperature.

Urea nitrate being very much less soluble at a low temperature than at the ordinary one is given an opportunity o crystallize out.

The mass of crystals is then thrown on an asbestos filter and drained by means of a vacuum pump and then washed several times with strong, well cooled nitric acid, to remove chlorides, phosphates and traces of other substances without dissolving much of the urea.

The crystals are transferred from the funnel to a beaker and dissolved in a very small quantity of boiling water. The solution is reprecipitated by adding pure cold concentrated nitric acid, and the crystalline and nearly pure nitrate now obtained is thrown on the same asbestos filter and drained thoroughly with the pump. The crystalline mass is dissolved in a small quantity of warm water and to the solution is added barium carbonate in slight excess, which forms barium nitrate with liberation of urea. The carbonate is added slowly as long as gas is given off on shaking and then a little excess. The mixture is evaporated to dryness on a water-bath and then treated with pure, strong alcohol, which dissolves out the urea but leaves the barium nitrate and excess of barium carbonate.

The alcoholic filtrate so obtained is usually colored. It may be purified by heating it gently with washed animal charcoal and filtering again. The final filtrate is evaporated at a low temperature and then allowed to crystallize spontaneously in the form of short needles.

By careful work pure urea can be obtained by the above process, but more readily by the synthetic method next to be given. This method is interesting as being essentially the one by which it was first demonstrated that organic compounds can be built up synthetically in the laboratory.

b. From Ammonium Cyanate. Fuse 100 Gm. of high grade potassium cyanide at a low red heat and oxidize it to cyanate by adding red lead, a little at a time, with constant stirring, while the mass is kept liquid. About 500 Gm. in all should be added. When the reaction is complete allow the mass to cool, powder it and extract with cold water. Treat the filtrate with solution of barium nitrate as long as a precipitate (barium carbonate) forms and filter again. The solution now contains potassium cyanate and several impurities. Add to it a slight excess of solution of lead nitrate which throws down a fine white precipitate of lead cyanate. Allow it to settle, wash by decantation and then on a filter with cold water, and finally dry at a low temperature.

By heating equivalent molecular weights of the pure lead cyanate and ammonium sulphate with a small quantity of water lead sulphate is precipitated and ammonium cyanate formed, which, however, on evaporation is transformed completely into urea. Extract the evaporated mass with absolute alcohol and crystallize as before.

The experiment is also made by starting with potassium ferrocyanide, which is dehydrated by heat and then fused with manganese dioxide to oxidize it to cyanate. The crude potassium cyanate is extracted by water and then the solution is treated directly with ammonium sulphate. On evaporating to dryness urea may be separated from the potassium sulphate by strong alcohol.

Pure urea melts at a temperature of $133°$ C. When heated to $150°$ a large part of it is converted into biuret, which can be recognized by the test already described. Its solution treated with sodium hypochlorite yields free nitrogen, carbon dioxide and water, while with nitrous acid a very similar decomposition is produced. The nitrogen of the acid escapes here, too.

These reactions form the basis of a method for the determination of urea, which will be explained in the section on urine analysis. Urea heated with glycocoll (amido-acetic acid) to about $200°$ C. yields uric acid but the reaction is probably devoid of physiological importance unless it can be shown that it can be brought about at a low temperature.

Problem 2. The Synthesis of Uric Acid.

The production of uric acid by the method just referred to may be illustrated by the following experiments.

We take up first the production of the glycocoll which can be accomplished by several methods, but most conveniently by decomposition of hippuric acid by means of sulphuric or hydrochloric acid by aid of heat.

Hippuric acid, which is the chief nitrogenous substance excreted in the urine of the herbivora, is in effect a combination of benzoic acid and glycocoll, or benzoyl glycocoll. The action of the aqueous acid in splitting it is one of dehydration as illustrated by this equation.

$$CO_2H.CH_2NH.(C_6H_5CO) + HOH = \\ CO_2H.CH_2NH_2 + C_6H_5CO_2H.$$

That is, benzoic acid and glycocoll are obtained as products. On the large scale the reaction is carried out in Germany in the commercial manufacture of benzoic acid.

a. To Prepare Glycocoll. Boil 10 Gm. of hippuric acid with 40 Cc. of dilute sulphuric acid (1 part of acid with 5 parts of water), ten to fifteen hours in a flask furnished with a return condenser for condensation of vapor. The benzoic acid separates as an oily layer, which solidifies on cooling. At the end of the boiling, while the liquid is still warm it is poured out into an evaporating dish and allowed to stand over night in order to let the benzoic acid settle out as completely as possible. The liquid is poured through a filter and the residue washed in water which is poured through the same filter.

The mixed filtrate is concentrated to a small volume which drives off some of the remaining benzoic acid. The rest is removed by shaking with ether in which the glycocoll sulphate is not soluble. The liquid is diluted with 200 Cc. of water, and neutralized by addition of barium carbonate, which is added in slight excess. A precipitate of barium sulphate separates. The liquid is decanted and the precipitate washed several times with boiling water, the washings are poured through a filter, and finally, with that poured off first, evaporated to a small volume. If the liquid is not clear it must be filtered before the concentration is carried far.

When reduced, finally, to a small volume the solution of glycocoll, now nearly pure is allowed to stand over night for crystallization. From the mother liquors several further crops of crystals can be obtained by concentration. The several products are mixed and purified by recrystallization from pure water.

b. The Formation of Uric Acid. Mix 1 part of glycocoll with 10 parts of pure urea and heat the mixture to a temperature between 200° and 230° C. The mixture darkens and becomes finally pasty and then hard. It is removed from the heat and allowed to cool, then broken up and dissolved in a boiling hot weak ammonia solution. An insoluble part is separated by filtration. From the filtrate the uric acid is precipitated by addition of a solution containing magnesia mixture and ammoniacal silver nitrate. The precipitate is washed several times with weak ammonia water, then stirred up with hot distilled water and decomposed by addition of solution of sodium sulphide. A precipitate of silver sulphide is filtered off. Hydrochloric acid is added in slight excess to the filtrate which is concentrated and allowed to cool. Uric acid crystallizes out in forms which can be recognized under the microscope. This crude product can be dissolved in weak ammonia, and precipitated again with the ammoniacal silver-magnesia mixture. By completing the process as before a purer acid is obtained which yields the tests described in the sections on urine analysis.

Uric acid can be obtained in quantity best from the excrement of birds or reptiles. It can also be precipitated from human urine. When pure it is perfectly white, but when thrown down from urine is always yellow or reddish.

Problem 3. Separation of Glycocholic Acid.

Some simple tests for the acids existing in bile have already been given. The separation of glycocholic acid in nearly pure condition is shown by the following experiment.

Glycocholic Acid. Evaporate 1000 to 1500 Cc. of ox bile to one-fourth its volume, then add some washed animal charcoal and evaporate with frequent stirring until the mass becomes practically dry. The whole evaporation must be carried out on the water-bath. The dry residue is rubbed up in a mortar then put in a flask on a water-bath and covered with twice its volume of absolute alcohol.

The flask is connected with an upright Liebig's condenser. By keeping the alcohol at the boiling temperature fifteen to twenty minutes the acids go into solution. The liquid is filtered and the mass left washed with warm alcohol.

If the mixed filtrate with washings is highly colored it must be heated again with fresh animal charcoal. The alcoholic solution is evaporated to a thick syrup. This is dissolved in a small quantity of absolute alcohol, the solution filtered and treated with absolute ether until a permanent turbidity appears. The mixture is allowed to stand in a closed vessel in a cool place until the sodium salts of the acids crystallize out in fine needle shaped crystals. These crystals are dissolved in a small amount of water and to the solution is added dilute sulphuric acid until it becomes turbid from precipitation of the free acid. On adding ether to the liquid the glycocholic acid separates in small needles on standing. These needles are pressed out and dissolved in hot water and the solution filtered. On cooling, the acid appears again. It may be made quite colorless by crystallizing from hot water several times. (*Drechsel.*)

It was shown by *Huefner* that a crystallization of nearly pure glycocholic acid may often be secured by adding hydrochloric acid and ether to bile directly. The method, as worked out by Dr. *J. Marshall* with American ox bile, is given here.

Measure out 100 Cc. of bile, add a few drops of hydrochloric acid, shake and filter without delay. Then add 5 Cc. of strong ether (or petroleum spirit) and then 30 Cc. of strong hydrochloric acid. Shake the mixture and allow it to stand a day, corked, in a cool place.

Crystals soon appear which are pressed out and recrystallized from hot water.

This reaction is not given by all biles, and to secure a good result it is necessary to use perfectly fresh bile. The acid and ether must be added within half an hour of the time the gall bladder was removed. The crystals are almost colorless when formed. Taurocholic acid is much more soluble and is not so easily obtained.

Problem 4. Preparation of Leucin and Tyrosin.

It has already been shown that in the prolonged action of the pancreatic secretion on albuminoids these two substances appear among the important end products. Their purification, however, when prepared in this way is a matter of some difficulty and other processes are resorted to when material for experimental purposes is desired. By the prolonged action of hot dilute acids on albuminous or kindred substances the two compounds are made in considerable quantity. Horn shavings are now frequently employed as the starting material.

Preparation. In a large flask boil 500 Gm. of horn shavings with 1,200 Gm. of thirty per cent sulphuric acid during ten to twelve hours, replacing evaporated water from time to time. Then neutralize the excess of acid with barium carbonate and finally with barium hydroxide in very slight excess. Filter through muslin, press out the precipitate, boil it with water and filter again hot. The mixed liquids contain a little baryta which is removed by careful addition of dilute sulphuric acid and filtration. Now evaporate the filtrates until a film of crystals appears, allow to cool, remove the crystals by filtration and evaporate again until crystals appear. Separate these and repeat the operations as long as crystals can be obtained. The first crystallizations consist largely of tyrosin, the last of leucin. The crude products are mixed and purified as follows: Dissolve in weak ammonia by aid of heat and add solution of basic acetate of lead as long as a precipitate forms and in slight excess, the solution being kept hot. Filter from the precipitate and through the filtrate pass hydrogen sulphide to remove lead in solution. Filter again and allow the liquid to cool. Usually, without further concentration, the tyrosin crystallizes out on cooling.

If crystals do not appear, concentrate slightly and allow to stand.

After the tyrosin has separated the mother liquor is concentrated by evaporation to a small volume. To this is added an excess of freshly precipitated copper hydroxide which gives a blue solution with leucin on boiling. Filter hot and concentrate the filtrate which precipitates a compound of copper and leucin. This is collected, mixed with water and treated with hydrogen sulphide, which precipitates the copper and leaves the leucin now in solution relatively pure. On concentration crystallization follows. The crystals can be purified further by dissolving in hot alcohol and setting aside in a cool place for a second crystallization.

As leucin is readily soluble in water its purification is by no means a simple matter. Under favorable conditions from 100 parts of horn shavings, 3.6 parts of tyrosin and 10 parts of leucin may be obtained. Other nitrogenous substances have been shown to yield even greater quantities of leucin. In general tyrosin is formed in much smaller amount than leucin.

Problem 5. The Use of the Spectrophotometer.

Under the head of blood tests the value of the spectroscope in the qualitative examination of blood was pointed out. In a modified form the instrument has become equally valuable in quantitative investigations and a short explanation of such applications will be given here. Used as a quantitative instrument the *spectroscope* is converted into a *spectrophotometer*, and forms have been devised by *Vierordt, Huefner, Glan, Wild* and others which render good service. Only one of these will be described here, the *Vierordt* form, as made by *Kruess*, of Hamburg.

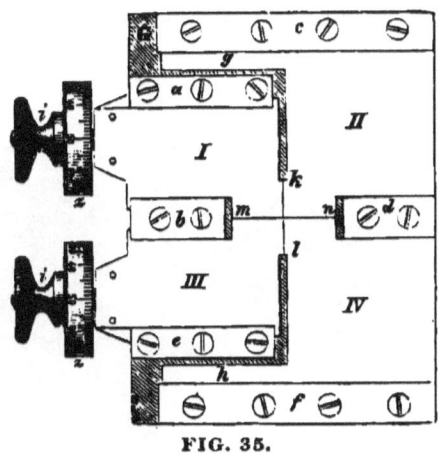

FIG. 35.

The essential features of this instrument are:

First, a double slit instead of the single slit of the ordinary spectroscope. The common slit is divided into two portions, an upper and a lower half, each controlled by its own micrometer screw. In the later instruments these halves open symmetrically, that is from both sides of a central line, instead of from one side as in the common Bunsen spectroscope. The width of each slit can be accurately measured by a micrometer screw. The construction and operation of the slit are shown in the figure above.

Second, the arrangement of the ocular tube by means of which a definite portion of the spectrum can be brought into the field of vision. The ocular tube can be given a lateral motion by means of a fine micrometer screw so that light of perfectly definite wave length can be brought before the center of the eyepiece at will.

In the eyepiece itself there is a movable framework which, when shoved to the left, brings two fine cross-hairs exactly in the center of the field and when shoved to the extreme right brings an adjustable slit in the same position.

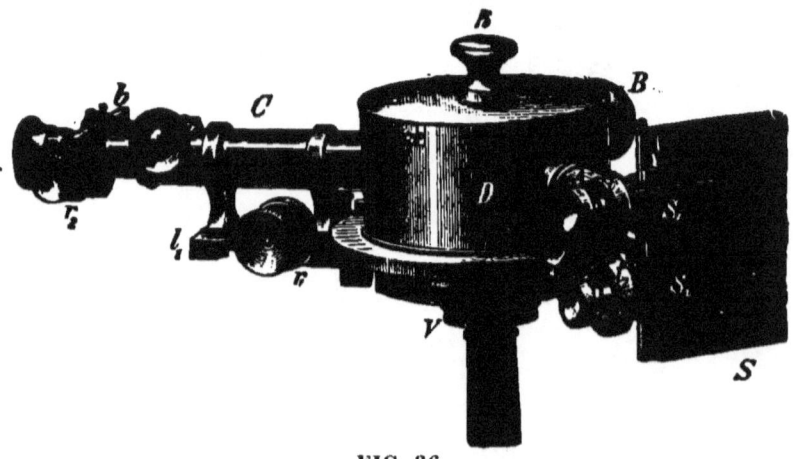

FIG. 36.

This slit can be opened or closed by a micrometer screw symmetrically, but its center has the position of the cross-hairs in the previous adjustment. Opening or closing this slit has the effect of taking in more or less of the spectrum.

In the above figure is shown the instrument as a whole and in the following one the measuring arrangements of the observation tube just referred to. C is the observation tube which is moved by the micrometer screw r_1, shown in detail in Fig. 37. The ocular slit is opened by the micrometer screw r_2, which moves also the cross-hairs, when necessary, and which is shown in detail in Fig. 37. Beneath and firmly attached to the observation tube is a scale l_1, which moves past the fixed pointer with mark at i_1. One revolution of the screw r_1 moves l_1 one division. The head of the screw is divided into 100 parts so that the position of the tube can be noted in scale divisions and hundredths, in four figures in all, as 2,852.

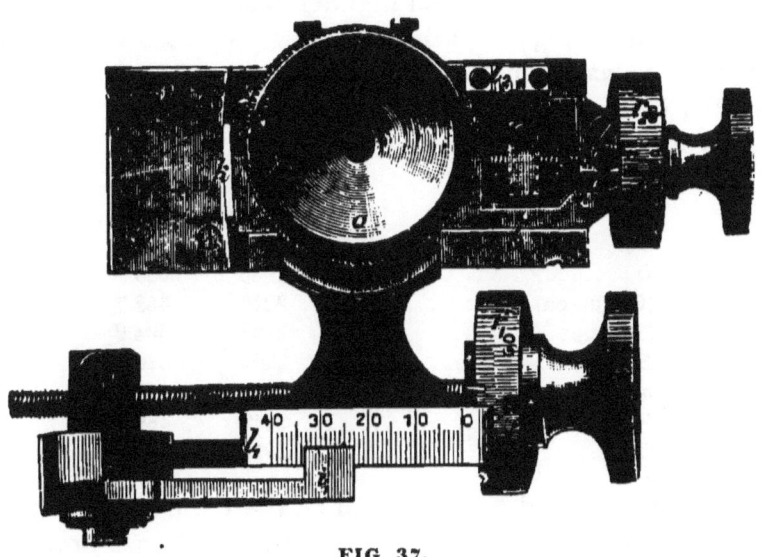

FIG. 37.

Suppose now with the ocular slide shoved to the left, that is, with the cross-hairs in the center of the field in perfectly definite position, the instrument is directed toward the sunlight and focused with the Fraunhofer lines sharply defined, it is plain that by moving r_1 any one of these lines can be brought to the center of the cross-hairs. When this is done the exact position of the tube can be read off on l_1 and r_1. If, with the D line, for instance, on the cross-hairs the position of the observation tube in scale division is 1,946, we know that always when we bring the tube in this position we have light of wave length 588.9 in the center of the field of view.

Before using the instrument for practical measurements it must be graduated by a method indicated by what has just been said. Sunlight is thrown directly into the collimator slit, S, and one by one the principal and characteristic Fraunhofer lines are brought to the center of the cross-hairs. The position of the observation tube for each line in the center of the field is then read off, giving finally a table connecting the arbitrary scale divisions with light of definite wave length. Expressing wave lengths in millionths of a millimeter a table like the following can be easily made:

Fraunhofer Line.	Scale divisions.	Wave lengths.
C	1750	656.2
D	1946	588.9
(Ca, green)	2079	558.7
(Tl)	2168	534.9
E	2203	526.8
b	2256	517.1
F	2446	486.0
G	2916	430.6

Between b and F, and F and G, there are many sharp lines which can be easily distinguished. These are brought also to the cross-hairs, and the corresponding position of the observation tube noted. Our table becomes extended so as to embrace many lines from the middle part of the spectrum. With this done we are able to bring any part of the middle spectrum under observation in the center of the field by simply moving r_1 to the corresponding position. If it is wished to examine light of wave length 534.9, that is light similar to the Thallium green, the observer moves the micrometer screw until l_1 and r_1 show the position 2,168. If light of wave length 546.7 is wanted, this being midway between the neighboring Ca and Tl greens, the observation tube is brought to the position 2,123.

For wave lengths near together this interpolation is satisfactory, but for greater differences it will not answer as the differences between the scale readings are not proportional, strictly, to the differences between wave lengths corresponding. Interpolations in longer stretches are best made from an interpolation curve, obtained by plotting the wave lengths as ordinates, and the scale divisions as abscissas. (The scale divisions given in the above table, as illustrations, are for a particular instrument only. The divisions corresponding to certain wave lengths would differ in different instruments.)

In order, now, to bring into the center of the field of view a certain color only, the ocular slide is shoved to the right, bringing the slit k in the center instead of the cross-hairs. This slit k may be made narrow or wide by motion of the micrometer screw r_2, but its center keeps the position which the cross-hairs formerly had and the light coming through it has the mean wave length of that corresponding to the scale divisions l_1, r_1.

As the ocular slide, and consequently the cross-hairs, can be moved by the screw r_2 it is possible to express the width k in wave lengths by this procedure. With the cross-hairs in position in the eyepiece bring a clearly defined Fraunhofer line to the center. Meanwhile, r_2 and l_2 must be at zero position. Then, by means of the screw r_1 the line is moved away from the center of the hairs, say fifty divisions on r_1. Next, by means of r_2 bring the line and center of hairs to coincide again, and note how many divisions r_2 must be moved through to do this. If a certain motion of r_2 moves the cross-hairs here through a distance corresponding to x scale divisions, the slit k would be opened to correspond to twice this number of divisions, because its two sides move symmetrically and only *one* is moved parallel with k. These adjustments are very carefully made on the instrument and enable the observer to limit with great exactness the extremes of wave length brought into view.

With the explanations here given other details can be learned from a study of the instrument itself. We turn next to a discussion of the principles involved in its use.

It was shown in the simple spectroscopic tests of blood in different degrees of dilution that the amount of light absorbed in passing through the liquid is closely dependent on the concentration, or in other words on the amount of hæmoglobin present. Strong solutions are nearly opaque. With increasing dilution more and more light comes through. Finally, with oxyhæmoglobin, much of the light between the red and blue is allowed to pass except in two narrow regions. Here two dark bands still show a strong absorption. If the dilution is carried still further these bands grow fainter but their positions remain unchanged. The points of maximum absorption are constant and characteristic for each substance.

Something similar is shown with solutions of potassium permanganate. If we observe these in cells with parallel walls, we find, for instance, with a certain dilution a heavy absorption band between the lines E and F, growing fainter from F toward a line with wave length λ 470.2. On dilution of this solution the band observed begins to break up, when finally we can easily distinguish 5 bands between D and a point a little beyond F (λ 588.9—λ 486, where λ stands for wave length). With increased dilution the bands grow fainter and the spaces between them wider and brighter. There must be, therefore, some simple relation between the amount of light absorbed and the concentration of the absorbing solution, or in other words, the number of molecules of coloring substance brought between the source of light and the slit. *If we know this relation and can measure the extent of the light absorption* it is evident that we have a means of arriving at the amount of the absorbing substance in solution.

Light passing through any medium, air, glass, water, or colored liquids, suffers a certain diminution in intensity. A certain amount of it is absorbed, this varying with the constitution of the medium, but following a very simple law for different concentrations or thicknesses of the same medium.

Increasing the thickness of the layer or column of absorbing substance has the same effect as increasing the number of absorbing molecules. To double the thickness of the cell of blood between the lamp and spectroscope slit amounts to multiplying by two the number of oxyhæmoglobin molecules which exert an absorbing action on the light.

But it does not follow from this that increasing the thickness of the blood layer or the number of molecules will diminish the intensity of light passing through in the same proportion. The absorption of light in various media follows a different general law which was first worked out by *Lambert*, on the assumption of a variation in the thickness of the medium rather than of its concentration. The relations found by *Lambert* for glasses of different kinds and which have been shown by later physicists to hold good for liquids, may be indicated by the following.

Suppose we have a source of light of intensity, I, and allow it to pass into a substance, which we will assume is divided into layers. In passing through the first layer, that is passing a certain number of molecules, its intensity is reduced to $\frac{1}{n}$ or amounts to $I.\frac{1}{n}$.

In passing through the next layer of same thickness it is again reduced by loss of the same fraction and is now

$$I.\frac{1}{n} \cdot \frac{1}{n} \text{ or } \frac{I}{n^2}.$$

In passing the next layer it becomes

$$I \cdot \frac{1}{n} \cdot \frac{1}{n} \cdot \frac{1}{n} = \frac{I}{n^3},$$

and in passing p such layers it becomes

$$\frac{I}{n^p}.$$

The remaining intensity of the light after passing p layers of absorbing substance may therefore be expressed by the formula

$$I' = \frac{I}{n^p}. \qquad (1)$$

This relation holds good only for homogeneous light or for light from a small region in the continuous spectrum.

For purpose of further calculation it may be assumed, arbitrarily, that the orignal light had the intensity $I = 1$. Our formula then becomes

$$I' = \frac{1}{n^p} \qquad (2)$$

from which

$$\log I' = - p \log n \qquad (3)$$

or

$$\log n = - \frac{\log I'}{p} \qquad (4)$$

The light absorbing power of two substances can be compared by noting the thickness of layers of these substances which must be taken to reduce the incident light to a certain fraction of the original intensity after passage. If a certain solution of blood coloring matter through which light passes has a thickness of 20 Mm., and if of another blood solution a thickness of 40 Mm. must be taken to reduce the transmitted light to the same fraction it is evident that the first solution must have a much greater light absorbing power than the second.

In comparing solutions or transparent solid substances practically it may be agreed to consider a reduction of the intensity of the light to $\frac{1}{10}$ its original value as the basis of comparison, and the shallower the layer of substance required to bring about this reduction to $\frac{1}{10}$ the greater must be its light absorbing power. The light extinguishing power of a substance or its *coefficient of extinction* has been defined by *Bunsen* and *Roscoe* as *the reciprocal value of the thickness of a layer of the substance necessary to reduce the intensity of the transmitted light to $\frac{1}{10}$ its original value.*

Representing the extinction coefficient by E and the reduced intensity by I' we have from the above formulas

$$E = \frac{1}{p} \text{ and } I' = \frac{1}{10}$$

$$\text{and } \log n = -\frac{\log \frac{1}{10}}{\frac{1}{E}} = E \tag{5}$$

Therefore

$$E = -\frac{\log I'}{p} \tag{6}$$

If p is given, once for all, the thickness of unity, that is if $p = 1$ (as 1 Cm.), our formula becomes

$$E = -\log I'. \tag{7}$$

It was said above that increasing the thickness of a layer of absorbing substance has the same effect as increasing its concentration in the same degree. From this it follows that the extinction coefficient must be *directly* proportional to the concentration. If E and E' represent the extinction coefficients and C and C' the concentrations of two solutions of the same substance

$$E : C :: E' : C'$$

represents their relation.

The further relations

$$\frac{E}{C}, \frac{E'}{C'}, \frac{E''}{C''}, \text{ etc.}$$

must be equal and constant, and must serve as *a characteristic connecting the light absorbing power of a solution with its strength.*

The term *absorption rate* has been applied to the ratio C : E by *Vierordt* and designated by him by A. Therefore

$$\frac{C}{E} = A$$

In illustration of this suppose we prepare a permanganate solution containing 0.25 Gm. of the pure crystals in a liter. Its concentration is, therefore, 0.00025. Suppose we further find by proper examination that the intensity of the light after absorption in a cell through which it is directed is 0.0436, a little more than $\frac{1}{25}$ of its original value.

From the formula

$$E = -\log I' = -\log 0.0436 = 1.36051$$

we have

$$A = \frac{0.00025}{1.36051} = 0.000184.$$

Now, experiment shows that for light *of constant wave length,* ($\lambda 494.7 - \lambda 486.5$ in the above illustration,) A is a constant for all strengths of solutions of the same substance. Finding A, once for all, in a definite region of the spectrum, we can use the formula

$$C = E \cdot A$$

to find the strength of a solution of which, experimentally, we are able to determine the extinction coefficient. Quantitative spectrum analysis by absorption is based on these principles.

It remains, now, to explain the methods for the determination of the extinction coefficient by means of the apparatus described above. This is most commonly done by aid of the divided slit of the spectrophotometer. The liquid under examination is placed in a cell with parallel, clear glass walls, 1 Cm. apart. The cell is half filled and so placed before the slit that the surface of the liquid is even with the dividing line between the upper and lower halves.

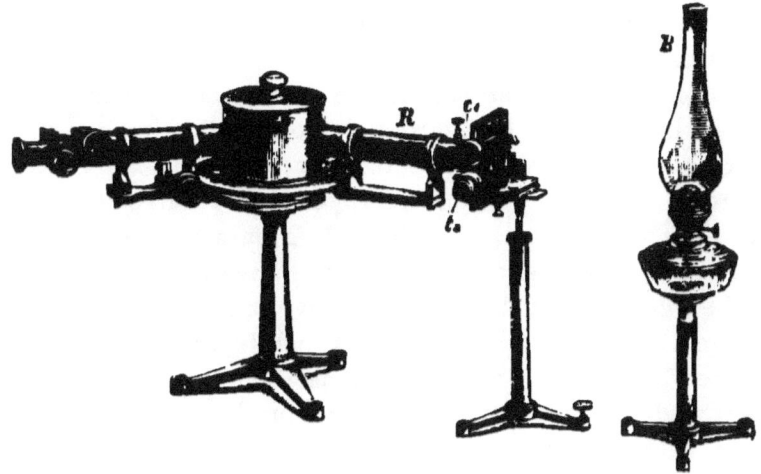

FIG. 38.

The light from an oil lamp is allowed to shine through the cell into the instrument, the edge of the flame being directed toward the slit and its center being just level with the center of the latter and in a line with the center of the collimator tube. This position of the lamp is essential to uniform illumination. It may be placed from 1 to 2 decimeters from the cell, which in turn must be as close to the slit as possible. The general arrangement of the apparatus is shown above.

With this arrangement light from the lamp enters the upper slit through the air and the lower one through the solution, which absorbs a part of it. If, before the solution is placed in the cell, the upper and lower halves of the divided slit are opened to the same width, which is shown by the graduation on the micrometer screw head, the projection of these in the eye piece will have exactly the same illumination, which can be best seen by cutting out all but a small part of the spectrum as explained above. With the cell, half filled, in position, the light coming through the lower slit is much weakened. The intensity in the two fields can be restored by opening the lower slit or closing the upper one to some extent. A better plan is to give the lower slit a definite standard width to begin with, say that corresponding to one whole turn or 100 divisions on the micrometer screw. Then the upper slit is narrowed until the intensity of the light allowed to pass is equal to that through the lower slit and solution. The reduced intensity of the light, or I' of the formula above, is given by the reading on the upper micrometer. If while the lower slit is opened by 100 micrometer divisions, the upper one must be narrowed down to 18 divisions to bring the fields to the same intensity, this shows that only 0.18 of the light entering passes through the solution.

$$I' = 0.18, \text{ and } E = -\log. I'.$$

This, therefore, gives us E.

The method as outlined is, however, not very convenient because of the difficulty in adjusting the two fields in the instrument. The meniscus unavoidably formed in the partly filled cell projects a broad, dark band across the spectrum, which effectually prevents exact comparison of shades.

But this difficulty has been, in a great measure, overcome by the use of a device suggested by *Schulz*. He gives the cell a width of 11 millimeters instead of 10, and drops in its lower half a block of clear glass with two faces parallel and just 10 millimeters apart.

This rectangular block or prism of glass has such dimensions and is given such a position in the cell that all light entering the lower slit of the spectrophotometer must pass through it and through one millimeter of liquid besides. The cell is *filled*, not *half filled* as before, and all light entering the upper slit must pass through the 11 millimeters of liquid. This is accomplished by having the upper surface of the glass prism perfectly horizontal and

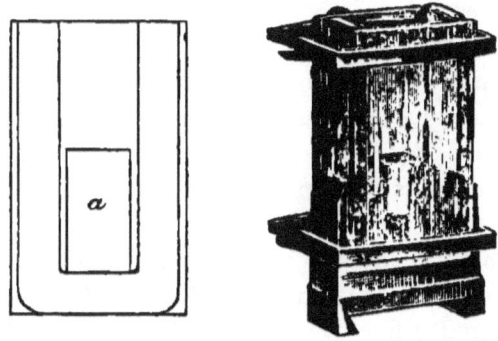

FIG. 39.

on an exact level with the dividing line between the two shutters of the upper and lower slits.

The above figure represents a cell with its rectangular glass prism, or *Schulz's prism*, in position.

In practice the cell is placed on a platform on the top of a standard, which by aid of a screw can be raised or lowered at will. The standard is attached to a solid iron base having a leveling screw, which is of value in bringing the Schulz's prism into proper position.

When the adjustments are properly made the projection of the upper glass surface appears as a line coinciding with the dividing line between the upper and lower spectra. No broad dark band appears to separate these two spectra to such a degree that their comparison is not readily made. The lamp is placed as before, but it is now the *upper* slit which is opened to the normal width, while the *lower* one is closed until the two fields become the same in intensity. The method of reading or making the calculation is not altered by the fact that the cell has now a width of 1.1 centimeters instead of 1, because in its lower part the light must pass through 1 millimeter of liquid as well as through the clear glass. The light absorbed by one millimeter below is equivalent to that absorbed by one millimeter of solution above, and the absorption of the remaining 10 millimeters above, alone comes into comparison.

In practice substances are dissolved in water, alcohol, ether, or other liquid for observation, and preliminary to actual tests the absorption relations of the solvent and the glass prism must be determined.

To do this, fill the cell with the clear menstruum, place it in position and examine with the lower slit opened to twenty or twenty-five divisions. Then adjust the width of the upper slit until equal intensities are secured. With water in the cell, and the Schulz prism of the usual glass, it has been found that with the lower slit at twenty-five divisions, the upper one must be brought to about 22.6 divisions to give the same intensity. The experimenter should make these tests for himself, however, with each new solvent used and with each Schulz glass body, as they are not absolutely uniform in their absorption power.

The absorption ratios of a number of physiologically important substances, for certain regions in the spectrum have been determined by *Vierordt, Huefner* and others and placed on record. The following table contains some of these results.

Spectral region.	Name of Substance.					
	Oxyhæmo-globin.	Hæmoglobin.	Methæmo-globin.	COhæmo-globin.	Bilirubin in chloroform.	Biliverdin in alcohol.
λ569.3—λ555.5...	0.00133	0.00122	0.00260	0.00131		
549.9— 540.0...	0.00100	0.00150	0.00199	0.00115		
558.1— 534.3...					0.00113	0.000215
501.2— 494.3...					0.0000598	0.000142
494.3— 486.1...					0.0000356	0.000116
486.1— 480.6...					0.0000209	0.000102
480.6— 474.4...					0.0000148	0.0000842
474.4— 468.4...					0.0000126	0.0000700
468.4— 461.7...					0.0000118	0.0000667

In this table the absorption ratios are mean values of several determinations made with solutions of different concentrations.

Problem 6. The Determination of Nitrogen.

The determination of the amount of nitrogen in an organic compound is a problem of the highest importance, and fortunately one of no great difficulty. Many processes have been in use for the purpose, some of the earlier ones requiring elaborate apparatus. These will not be described here. At the present time chemists apply the method of *Kjeldahl* in the great majority of analyses, and this alone will be explained.

The Kjeldahl Process. This is founded on the following facts. Nearly all organic compounds containing nitrogen, and probably all with which the physiologist has to deal, are decomposed when heated with strong sulphuric acid. If with the acid certain oxidizing agents are used the nitrogen becomes wholly converted into ammonia. As originally carried out by Kjeldahl and others, the substance is heated for an hour or more in a thin glass flask with sulphuric acid to a temperature near the boiling point of the latter. In this heating the organic substances decompose, and the solution becomes nearly colorless. Then, while still hot, powdered potassium permanganate is added in small amount, until a permanent green color is obtained in the oxidized liquid. It is then allowed to cool, diluted with water, neutralized with pure sodium hydroxide in excess and then distilled. The ammonia formed is caught in standard acid and measured as explained below.

The original process has been modified in several directions, and mainly so as to shorten the time required in oxidation. Kjeldahl himself suggested the use of strong fuming acid and the addition of phosphoric anhydride. Others have used certain metallic salts, and also mercury as an addition, with the result of reducing the time of digestion to less than an hour. More recently the process has been modified by *Gunning* in a manner which constitutes an improvement for most purposes. He found that the oxidation is much more perfectly carried out if to the sulphuric acid about half its weight of pure potassium sulphate is added, and that the subsequent addition of permanganate is then rendered unnecessary. Some few nitrogenous bodies are not decomposed by this mixture, but all that we have to consider here are completely oxidized.

The process is conducted in the following manner: Weigh out a gram or more of the substance to be analyzed. If the substance is in moist condition or in solution, concentrate nearly to dryness, after addition of a little pure sulphuric acid, in a platinum dish, or best in the flask in which the digestion is afterward made. If the solution is very weak it may be evaporated in platinum first, and then transferred to the digestion flask for the final concentration. To the known amount of the nearly dry substance in the digestion flask add 15 Cc. of *pure* strong sulphuric acid and 10 Gm. of *pure* potassium sulphate. Place the flask on a sand-bath or gauze, and heat with the Bunsen burner, gently at first, and afterward to a high temperature, until the liquid ceases to froth and becomes colorless or pale yellow. Round bottom, Bohemian flasks with long necks are best for this purpose, and they should be supported on the gauze in an inclined position to prevent loss by spirting.

The heating may last from half an hour to two hours, the time depending on the nature of the substance and its amount. The liquid in the flask is allowed to cool, and is then diluted with about 200 Cc. of water. The mixture is poured into a distillation flask, and to it a few drops of phenol phthalein solution are added. Then enough 50 per cent sodium hydroxide solution is added to more than neutralize the sulphuric acid present and liberate the ammonia. This alkali solution must itself be perfectly free from ammonia. Add some small bits of ignited pumice stone (about half a gram of fragments which will pass through a sieve with sixteen meshes to the linear inch) to the flask, connect with the condenser and distill, while the further end of the condenser dips beneath the surface of a measured volume of dilute standard sulphuric acid to catch the ammonia.

The distillation apparatus for that purpose should have the following construction: The flask should contain from 500 to 700 Cc., and should be closed with a rubber stopper through which passes a glass tube with an internal diameter of at least 12 Mm. This glass tube extends 5 Cm. or more into the neck of the flask and is cut off diagonally. Above the flask it should have a vertical length of about 30

Cm., in the middle portion of which it is widened to a bulb of about 3 Cm. diameter. Following the vertical part the tube is bent to give a horizontal length of 15 to 20 Cm., and then is bent down. This downward limb passes through a rubber stopper into the wide part of the condensing tube of a short, upright Liebig, or a spiral condenser. The lower end of the condensing tube dips beneath the surface of the weak acid which catches the distilled ammonia. The glass tube which leads from the distillation flask is continuous until the condenser is reached. With this arrangement there is no danger of losing ammonia or of carrying the fixed alkali over mechanically.

As standard acid it is convenient to use one-tenth normal sulphuric acid, which is colored with a *single drop* of dilute methyl orange solution, as indicator. The distillation is continued until about 150 to 200 Cc. has passed over into the standard acid, which must still show a pink color. The excess of acid is then titrated with weak standard ammonia, and the amount of acid found subtracted from that originally taken shows how much was required to combine with the ammonia in the distillation. 17 parts of ammonia, NH_3, correspond to 49 parts of sulphuric acid, H_2SO_4. For each 17 parts of ammonia found we calculate 14 parts of nitrogen, and thus learn the amount of this element in the weight of substance originally taken for analysis. As dry albuminous substances contain in the mean about 16 per cent of nitrogen, and if we know that we are dealing with a body of this class, we obtain the amount of albumin by multiplying the nitrogen found by the factor, 6.25. In illustration, if 50 Cc. of standard sulphuric acid, with 4.9 Gm. to the liter, had been measured out in a certain experiment to catch the ammonia, and if at the end of the distillation 18 Cc. of tenth normal ammonia solution is found necessary to neutralize the excess of acid it shows that the amount of ammonia distilled over is 32×1.7, or 54.4 Mg. This is equivalent to 44.8 Mg. of nitrogen in the original substance, which multiplied by the factor, 6.25, gives .280 Gm. as the corresponding amount of albuminoid.

It is not always possible to obtain sulphuric acid or potassium sulphate absolutely free from ammonia. In this case it is necessary to make a *blank* experiment, using the acid mixture alone, and determine the amount of ammonia which can be distilled from it in the usual way as described. This amount can then be subtracted from that found in the actual test.

The addition of pumice stone, above recommended, prevents bumping in the distillation of the heavy liquid. Various other substances have been recommended, but this is probably the most satisfactory as with it the liberation of vapor is uniform. The 150 or 200 Cc. of distillate may be collected in an hour.

Problem 7. The Separation of the Proteids, as Illustrated by the Analysis of Meat Extract.

Some of the characteristic reactions by which the important proteid compounds may be separated or recognized have been given in a former chapter. These, with others, may now be employed in systematic order for the quantitative analysis of a complex mixture such as we find in products of digestion or in meat extract. A general method to be followed in such cases has been given by *Stutzer*, and this will be applied here, with slight modifications only.

The Analysis. If the substance for examination is in dry, or nearly dry form weigh out 5 Gm.; if in paste form take 10 Gm., and if a liquid take 25 Gm. Dissolve it in about 150 Cc. of lukewarm water, and if a part remains insoluble, filter through a weighed Gooch crucible, wash the residue with distilled water and make the filtrate up to 500 Cc. The residue may be dried slowly at a low temperature and then at 105° C., and weighed, or while still moist, it may be washed into a flask with a little water, concentrated again and then analyzed for nitrogen according to the Kjeldahl method. Calculate the nitrogen to albumin, as explained in the last problem.

If the product contains fat this will appear as an oily layer on the water in which it is dissolved, and after the filtration will be found in the contents of the Gooch crucible. By drying the crucible, extracting with anhydrous ether, evaporating and weighing the residue the fat is determined.

We have now to examine the aqueous filtrate of 500 Cc., which may contain soluble albumin, albumose, peptone and other compounds of nitrogen of less complexity. Measure out 100 Cc. of this filtrate, add to it two or three drops of acetic acid and heat to boiling on wire gauze. If soluble albumin is present it appears as a coagulum, which may be collected on the Gooch crucible, dried and weighed, or it may be transferred to a digestion flask and be treated by the Kjeldahl method for albumin.

The filtrate and washings from the soluble albumin are concentrated to a small volume, about 10 Cc., and when cold treated with 100 Cc. of a cold saturated solution of pure ammonium sulphate. As has been explained this reagent precipitates albumose, but not peptone. The precipitate is allowed to settle and is then collected on a 10 or 12 Cm. filter and thoroughly washed with saturated ammonium sulphate solution. The filter paper used for this purpose must be free from soluble nitrogen compounds. Munktell's washed Swedish paper answers very well.

The albumose precipitate is next treated on the filter with lukewarm water, the solution running through being collected in a clean flask. Make the filtrate up to 250 Cc. It contains the albumose and also some ammonium sulphate. In order to determine the albumose concentrate 150 Cc. of this filtrate to a small volume and find its nitrogen by the Kjeldahl method. A part of this nitrogen, however, comes from the ammonium sulphate, and its amount must be found and deducted. This may be done, most conveniently, in the following manner. Take the remaining 100 Cc. of the albumose filtrate, add to it a few drops of hydrochloric acid and heat to boiling on gauze. Add now a slight excess of solution of barium chloride and boil some minutes longer. We obtain here a precipitate of barium sulphate which corresponds, of course, to the ammonium sulphate and consequently to the nitrogen of its solution. This precipitate settles well and can be readily collected on a Gooch crucible in the usual manner. Weigh it and reduce the barium sulphate found to corresponding nitrogen. Subtract this nitrogen, calculated for 150 Cc., from that found in the Kjeldahl test. The remainder is that due to the albumose. To obtain the latter multiply by the factor, 6.25, and calculate the amount for the 250 Cc. of filtrate. This gives us now the albumose in 100 Cc. of our first filtrate, or in one-fifth of the substance originally taken.

Of the important nutritious bodies found in meat extract, peptone remains to be determined. No reagent is known which precipitates this leaving the other proteids, hence the following method has been suggested by *Stutzer*. He finds that while peptone, albumose and albumin are completely precipitated by phospho-tungstic acid and are insoluble in excess, kreatin and kreatinin, precipitated at first, dissolve when more of the reagent is added. Leucin tyrosin, urea, and taurin, are not precipitated at all. Xanthin and hypoxanthin behave as do the proteids, that is, they form with the reagent insoluble precipitates. They are only slightly water soluble, however, and can be present in meat extracts or similar products in but small proportion. For our purpose their presence may be neglected. Therefore, as appears from what has just been said, if we precipitate peptone, albumose and albumin with phospho-tungstic acid, determine the total nitrogen in the precipitate, and from this subtract the nitrogen of albumose and albumin found by other methods we have left the peptone nitrogen, which, multiplied by 6.25 gives approximately the peptone.

To find the peptone in the case before us, we use the aqueous filtrate obtained after filtering out the insoluble albumin and other matters. Take 50 Cc. of this filtrate, acidify it strongly with sulphuric acid and add to it an excess of a reagent made by dissolving 100 Gm. of crystallized sodium tungstate and 25 Gm. of glacial phosphoric acid in 500 Cc. of water, to which afterward enough sulphuric acid is added to give a strong acid reaction. By this treatment the proteids, with traces of xanthin and hypoxanthin possibly, precipitate, while other nitrogenous bodies are left in solution. Allow the mixture to stand some hours and then filter it through a nitrogen-free filter paper. Wash the precipitate with the reagent and finally with a little dilute sulphuric acid. Allow it to drain and then transfer, with the filter paper, to a digestion flask and treat with the Kjeldahl oxidizing mixture in the usual manner. The nitrogen of albumin and albumose have been already determined. We subtract this from the nitrogen of the last determination and calculate the remainder to peptone by multiplying by 6.25, as the percentage composition of peptone is practically the same as that of the other proteids.

The water in the original sample may be determined by weighing about 2 to 5 Gm. into a small platinum dish and heating on a water-bath until constant weight is obtained. The loss in weight represents the water.

The mineral matter or ash may be found by incinerating the above dry residue and heating it until it becomes colorless. There is usually a small loss by volatilization here.

By adding together the water, mineral matters and proteids, and subtracting from 100 per cent, we have a remainder which represents the percentage of nonproteid nitrogen bodies. If fat is present it must be subtracted too.

A further idea of the amount of these nonproteid nitrogenous matters may be obtained by making a determination of total nitrogen in the original substance by the Kjeldahl method. By subtracting from this total nitrogen that of the proteids, found as just explained, we have the nitrogen of the kreatin, xanthin and other bodies. This multiplied by 3.12 gives, *approximately*, the amount of these substances because they contain in the mean about 32 per cent of nitrogen.

Part II.

Urine Analysis.

Chapter X.

OUTLINE OF TESTS. PRELIMINARY TESTS.

THE importance of an accurate knowledge of the bodies excreted by the urine has long been recognized and elaborate investigations have been carried out to determine the nature and quantities of these substances, some of which appear normally in health, while others are found only during the progress of disease.

Experiment shows that normally certain products occur in the urine in relatively large amounts, and give to it its prominent characteristics, while of other products the amounts present are so minute that their detection is a matter of no little difficulty.

Certain grave disorders are accompanied by the appearance of certain substances in the urine, and where the chemical or microscopic tests for the latter are simple and unquestionably correct we have at hand a convenient aid to diagnosis. In many cases, however, it is true that we are unable to trace the relation between small amounts of substances occasionally appearing in urine and any specific disorder or condition of the body. The detection of such substances is naturally without value in diagnosis, at the present time.

Yet it would be unwise to neglect the study of such traces because, as medical science progresses, new relations are from time to time brought to light which give value to data which at one time may have been considered wholly unimportant. Complete handbooks on the urine give prominence to many topics which will not be touched upon in what follows because we are here concerned with phenomena everywhere recognized as important and the bearings of which, in the main at least, are understood.

In the practical analysis of urine such as is customary for clinical purposes comparatively few tests are required and little apparatus is necessary beyond that already used for other examinations. Frequently a single test is sufficient to determine all the physician needs to know, for instance, regarding the presence or absence of sugar or albumin.

In the following pages those tests and processes will be described which have been shown by experience to be amply sufficient for all practical requirements. Some of these are qualitative, others quantitative and may be tabulated as follows:

Qualitative tests.
1. Observation of color and odor.
2. The reaction, whether acid or alkaline.
3. The tests for albumin.
4. The tests for sugar.
5. The tests for the characteristic biliary acids and coloring matters.
6. The various tests for blood.
7. Tests for other coloring matters.
8. The examination of the sediment.

Quantitative tests.
9. Determination of specific gravity.
10. " " the amount of albumin.
11. " " " " sugar.
12. " " " " uric acid.
13. " " " " urea.
14. " " " of phosphates.
15. " " " " chlorides.

The above includes the usual and important tests. A few

others will be given in the proper place, for instance, tests for acetone and diacetic acid, which under circumstances may have importance.

Normally, urine contains as its most important constituents urea, sodium chloride, certain phosphates and urates, and smaller amounts of other substances as hippuric acid, xanthin, kreatinin, traces of phenols, etc.

Pathologically there may appear albumin, sugar, blood, pus, bile pigments and acids, and a number of other bodies insoluble, or of slight solubility, which usually appear as a sediment.

We turn now to an explanation of the various preliminary tests employed.

Specific Gravity.

The density or specific gravity of the urine secreted in twenty-four hours, varies in health between rather wide limits, probably between 1.005 and 1.030. 1.020 may be taken as about the mean value at 15° C.

The specific gravity depends primarily on the amounts of liquid and solid food taken, and on the loss of water from the skin by perspiration. When this loss is great the specific gravity of the urine is correspondingly increased, other things being equal.

In disease the density may be lowered below or increased above the normal value.

For an absolutely exact determination of the density the use of the pycnometer, Mohr-Westphal balance, or other apparatus is necessary. But for our purpose the *urinometer*, or density bulb, is sufficiently accurate. This little instrument is shown in the following figure. The urine to be tested is poured into a narrow jar, about one Cm. wider than the bulb, and after the air bubbles have escaped, the urinometer is immersed in it. When it comes to rest the degree at which it stands is read off *below the surface*. Usually the last two figures only of the density are marked on the stem, as 25, instead of 1.025, and these are often given as the density.

As the density of urine decreases about one degree for an increase in temperature of 3° to 5° C. it is important that the test be made at a definite known temperature, as 15° or 25° C. Urinometers have usually been graduated to give the correct reading at a temperature of 15.5° C. (60°F.). But at the present time we have them for the temperature of 25° C. (77° F.), because this is with us a more common house temperature than the lower one. It is convenient to have the instrument indicate the correct specific gravity without the necessity of cooling. The specific gravity of urine varies approximately as does that of water, with changes of temperature. A table in the appendix shows the rate for water, and by the use of this a correction can be made.

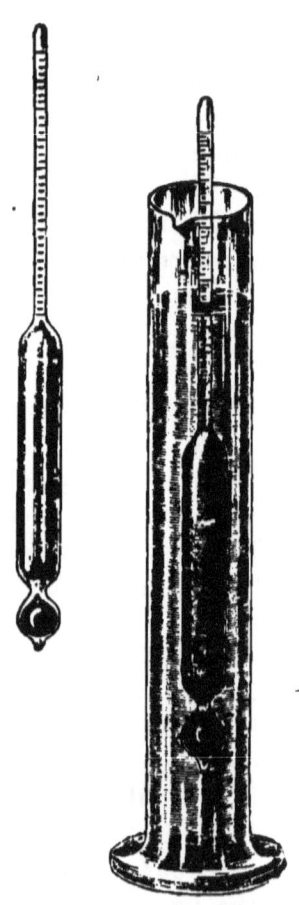

FIG. 40.

By noting the amount of urine passed in twenty-four hours, and the density of the mixed liquid, a rough determination of the solid matters contained in it can be made.

For this purpose it is simply necessary to multiply the last two figures of the density by 2.33 (known as the coefficient of *Haeser*), which gives the approximate number of grams in a liter. By proportion the amount for the day can be calculated from this.

For example, 1,400 Cc. of urine was passed, and its density was found to be 1.024,

Then, $24 \times 2.33 = 55.92$,

and, $1,000 : 1,400 :: 55.92 : x = 78.288$.

This calculation is frequently of service.

As indicated above a variation in the specific gravity of normal urine may be due to several causes, the most important of which are changes in the volume of water drunk or the weight of nitrogenous food and salt digested. The amount of urine excreted daily may be taken as 1,500 Cc. in the mean. Assuming that fifteen grams daily is the salt consumption and that it is all excreted with the urine we have through this factor alone a specific gravity of about 1.007. Assuming further that 150 grams of nitrogenous food (considered as pure albumin) is consumed daily and that four-fifths of this amount is daily excreted as urea, the weight of the latter in the urine would be forty-three grams. This alone would produce a density of nearly 1.008 and give a percentage composition of 2.84. Combined with the other substances in urine the relative effect of the addition of urea would be greater. All of the solids of the urine have a specific gravity greater than that of water and their presence therefore adds to the specific gravity of the excretion, but changes in the density, due to changes in the amounts of uric acid, phosphates, sulphates, etc., passed are of less importance because of the relatively small quantities of these substances normally present.

If, with the food consumed normal, the water taken is

small in amount, or if a large amount is lost from the skin as perspiration then the density of the excreted urine must be correspondingly higher. A large volume of water consumed or little evaporation from the skin will give a urine of lower density. It is plain, therefore, that great variations in the specific gravity of the urine may occur and from perfectly normal causes.

In disease even greater variations may occur, one of the most characteristic and important being that due to the presence of sugar in *diabetes mellitus*. Here the density may reach 1.040 or higher, while the volume of urine is above 1,500 or 2,000 Cc. in the twenty-four hours. A high specific gravity with *large volume* is always suspicious and suggests presence of dextrose although occasionally it may be due to presence of large doses of soluble salts taken into the system as remedial agents. A low specific gravity with *small volume* of urine must also call for investigation, as this points to the absence of or marked decrease in the normal constituents from some cause. A lower density is observed in diseases where the elimination of urea is slower because of hindered tissue changes, in conditions of malnutrition in general, and in any disease involving the structure of the liver itself. In *acute yellow atrophy* of the liver, for instance, urea is much diminished, and the specific gravity low.

The diminution in excreted chlorides, with normal consumption, in certain diseases is also a factor in causing low specific gravity. This may follow when the salt consumed is eliminated temporarily in various exudations or effusions rather than by the normal channel.

Some of these indications will receive attention later during the discussion of the tests for the common normal and abnormal urine constituents.

Reaction.

In health the reaction of the mixed urine passed through twenty-four hours is always acid.

This normal acid reaction is supposed to be due to the presence of acid phosphates and to small amounts of uric acid and to other free organic acids. The reaction can be observed by the aid of sensitive litmus paper, but the absolute amount of free acid is very small.

Occasionally urine is passed which gives the so-called *amphoteric* reaction with litmus; that is it turns blue paper red and red paper blue. It has not been found possible to connect this phenomenon with certainty with any definite pathological condition; it has, therefore, no special clinical significance at the present time. Some hours after a hearty meal an alkaline reaction is frequently observed, giving place soon to the usual acid condition. This alkaline reaction may be due to the presence of small amounts of trisodium phosphate formed during active digestion. The administration of alkali carbonates, or of certain organic salts, as malates, acetates, tartrates or citrates, which yield carbonates by final decomposition, may also occasion an alkaline condition. Sometime after it is voided urine always becomes strongly alkaline in reaction, but this change may be delayed for days or weeks even. It is brought about by the decomposition of urea, which is usually a result of bacterial action. In this decomposition ammonium carbonate is formed, the odor of which often becomes very strong. Anything which prevents or impedes the bacterial activity tends to maintain the ordinary acid or neutral reaction. Salicylic acid, thymol, chloroform, volatile oils and other antiferments behave in this manner, and are frequently added to specimens of urine to preserve them for investigation. For the preservation of 100 Cc. of urine one-fourth of a Gm. of salicylic acid is enough.

Sometimes the decomposition of the urea takes place in the bladder, the voided urine having then a very marked alkaline reaction and strong odor, usually. Such a change may be brought about by the progress of disease, or may be induced by the introduction of a dirty catheter into the bladder. This carries the organism capable of splitting up the urea, and the condition once established may be maintained for a long time.

Ammonium carbonate results from the reaction. This may be distinguished from the fixed alkalies (hydroxide or carbonate of sodium or potassium) by a very simple test. A piece of sensitive red litmus paper immersed in alkaline urine becomes blue. On drying the paper the color due to the nonvolatile alkalies persists, while that of ammonium carbonate disappears. The test has some practical value as it is necessary to distinguish between the alkalinity of urea fermentation and that of an excess of fixed alkali occasionally present. For these tests only fresh sensitive paper can be safely used. The conversion of urea into ammonium carbonate is represented by this equation:

$$CON_2H_4 + 2 H_2O = (NH_4)_2CO_3$$

In highly colored urines it is not always easy to observe the reaction with litmus paper. In this case the method described under the blood tests in an earlier chapter may be applied. This consists in immersing small discs of plaster of Paris in *neutral* litmus solution and then drying them. A few drops of urine are placed on a disc and allowed to remain some minutes. The urine is then washed off leaving a bluish or reddish spot indicating the reaction.

Odor.

The odor of urine is not easily described, as in health it is *sui generis* and characteristic. Normal urine contains traces of complex aromatic bodies the exact nature of which cannot in all cases be given. These substances are more abundant after a vegetable than after an animal diet, and are especially noteworthy in the urine of persons whose food contains such vegetables as cabbage, radishes, parsnips, asparagus or the spices. It is well known that certain substances given as remedies give rise to distinct odors in the urine. The administration of turpentine imparts to the urine an odor of violets.

As the odor so largely depends on the nature of the food it may be much modified even in health, and in disease may be characteristically changed. The ammoniacal odor of urea decomposition in the bladder has been referred to, and the peculiar *sweetish* odor of diabetic urine has long been noticed.

But it must be remembered that many strong odors may be developed in the urine soon after passage by the action of ferments other than the *micrococcus ureæ* which yields ammonium carbonate. In some cases these give rise to what may be called a putrefactive odor.

Color.

The color of urine is described as *straw yellow*. Many causes, however, may produce a change in this shade, leaving the urine still normal. As can be readily seen the color is closely dependent on concentration and must, therefore, vary with the amount of liquid taken into the stomach.

Certain foods from the vegetable kingdom possess characteristic coloring matters which pass, more or less changed, into the urine. As long as the latter is acid the presence of these may not be noticed, but with a change of reaction a change of color may follow, usually to reddish.

Santonin imparts a yellowish color to urine, reddened

by alkalies. In pathological conditions the color of urine is often characteristic and of great importance in diagnosis. The presence of blood, for instance, is indicated by a more or less sharp shade of red, bile by a peculiar greenish brown, especially noticeable in froth produced on shaking. The urine of *diabetes mellitus* is generally very pale, while the urine of fevers is usually highly colored not only from the diminution of water but also from the presence of abnormal coloring matters. Different shades are produced by the presence of *altered* blood and bile constituents, which will be referred to later. The real color of the urine is often obscured by loss of transparency due to precipitation. Normal urine is generally perfectly transparent when passed, but sometimes cloudy from presence of a suspended precipitate of mucus or phosphates. On becoming alkaline a precipitation of earthy phosphates usually follows. A precipitate of urates, without change of reaction, often takes place by simply lowering the temperature of the urine. This precipitate, however, disappears on the application of a slight heat to the urine, and leaves the latter clear for examination of color.

Clinically, the color indications of greatest importance are those due to the presence of derivatives from the blood or bile. These may be unaltered elements of the blood or bile or decomposition products of their essential coloring matters.

In a following section on the tests for abnormal coloring matters in urine this question will be again taken up.

Chapter XI.

THE TESTS FOR ALBUMINS.

ALBUMINOUS bodies do not occur in normal urine except, perhaps, in mere traces. Numerous investigations have been published on this subject, and while some of the recent ones would seem to show the probability of a *physiological albuminuria*, others, seemingly as thorough, lead to quite the opposite conclusion.

Temporarily, it is true, albumin may be found in the urine of healthy individuals, as after the consumption of large quantities of egg albumin, or after the action of some cause producing a sudden alteration of the blood pressure, but the amounts found in such cases are too small, and their occurrence too rare to permit them to be classed as anything but accidental. It is certain that the presence of any appreciable amount of albumin in the urine and the persistence of the same must be looked upon as a *pathological* phenomenon and one of the greatest importance to the physician.

Albumins may appear in urine from several sources, most frequently, probably, because of some structural change in the tubules of the kidneys which permits a filtration from the blood. But this is not always the case as they may appear from sources in no way dependent on renal disorder, from lesions of the ureters, bladder or urethra, for instance, in which case blood or pus may be present. Ordinary serum albumin is the usual, but not the only proteid body which may appear in urine. Half a dozen or more modifications have been described as occurring under different circumstances, but the evidence for some of these appears to be of doubtful character. Certain forms are not readily detected or identified. In what follows tests will be given for those proteid bodies which can be detected with certainty and whose presence has some definite clinical importance.

1. Serum Albumin.

The presence of serum albumin in the urine is a characteristic of what is ordinarily termed *albuminuria*. As intimated above albuminous bodies may appear in the urine from different sources. The presence of serum albumin suggests (*a*), a functional or structural disorder of some part of the essential tissue of the kidney, in which case we have *renal albuminuria* or *true albuminuria*, or (*b*), a lesion of some part of the urinary tract below the kidney, in which case we have what is called *false* or *accidental albuminuria*.

Renal albuminuria is the condition appearing in Bright's disease or *acute parenchymatous nephritis* and in other pathological conditions in which a change of the diffusion membrane is involved. It is also frequently induced by derangements in the circulation due to heart diseases, high fevers, etc., which in turn may react and give rise to a derangement of the kidney itself. That is to say, the causes producing certain febrile conditions may extend to the structure of the renal filtering apparatus and so alter its condition that the passage of albumin is no longer hindered but becomes continuous.

Under all such circumstances the albumin passing through the kidney is generally accompanied by something which suggests its origin. There may be here an excessive amount of the epithelial lining of the tubules, or plugs of coagulated albumin, mucus, or of the wax-like, partially degenerated albumin known as lardacein, all in the form of "casts" of the uriniferous tubules. These may be readily seen and recognized by the microscope.

As intimated above, false or accidental albuminuria can originate from several causes and in general is a condition of far less clinical importance than the other. It is usually possible to determine by a few examinations the real nature of the disorder by aid of the facts just mentioned.

Because of the very great importance of the subject to the physician, much attention has been given to the question of albumin tests, and the number of reactions proposed for its detection reach, possibly, the hundreds. Many of these are of such extreme delicacy and so easy of execution that to make a choice of a few is by no means a simple matter.

The best of them depend on the fact that the soluble serum albumin, which finds its way into the urine, can be coagulated and made visible as white flocculi, or as a white cloud when present in small quantity. Of the various methods of producing this coagulation, only those will be mentioned which are most characteristic, and practically the most useful.

QUALITATIVE COAGULATION TESTS.

Coagulation by Heat. When a sample of urine is boiled a precipitate usually forms. This in most cases consists of *earthy phosphates*, and is often sufficient to conceal a precipitation of albumin possibly present. If now to the boiled sample about one-tenth its volume of strong nitric acid be added, the precipitated phosphates will disappear, while the albumin will remain coagulated. It is necessary to add as much nitric acid as is here indicated, *because a small amount may sometimes dissolve coagulated albumin, forming soluble acid-albumin.* This acid albumin is broken up on the addition of more acid.

Even when boiling does not throw down a precipitate, the addition of nitric acid cannot be omitted, as under certain circumstances the heating may produce a soluble combination between alkalies present and albumin, which is stable. Nitric acid in sufficient quantity will break up this combination, and bring about coagulation.

Under most circumstances this heat test, as outlined, is sufficient, and the possibility of making a mistake is very small. It has been shown in an earlier section of the book that small amounts of albumin combine readily with weak acids and alkalies, forming soluble and stable combinations known as *acid-albumin* and *alkali-albumin.*

If the urine has a neutral or alkaline reaction to begin with a small amount of alkali-albumin would escape detection by heating alone. On addition of just the proper amount of acetic acid to neutralize the alkali, the application of heat will cause a coagulation, but a *slight* excess of this acid might convert the *alkali-albumin* into *acid-albumin,* equally hard to precipitate. Traces of nitric acid, and in a marked degree hydrochloric acid, behave in the same manner, but the addition of larger amounts of nitric acid is free from this objection because in proper amount this acid is able to decompose both acid and alkali-albumin.

When taken for examination, urine is frequently cloudy from the presence of precipitated urates or earthy phosphates. Heat is sufficient to dissipate the cloud if due to the urates, but the phosphate cloud is rendered heavier. It is always a good plan to carefully filter the urine, if in the least degree turbid, before undertaking the test.

With old samples of urine which have undergone the urea fermentation and have become alkaline, the test by heat and subsequent addition of acid is not always satisfactory or convenient. In such cases it is best to proceed at once to a method which disposes of the excess of alkali at the start and in such a manner as to cause no confusion.

Coagulation by Nitric Acid. As indicated above, nitric acid can coagulate albumin, and this test is frequently employed without previous boiling. When applied to fresh urine the test may be made in this manner.

Several Cc. of the strong acid are warmed in a test-tube, and over this is carefully poured an equal volume of urine, so as to overlie without mixing. If albumin is present a white ring appears at the surface between the two liquids. When the urine contains an excess of coloring matter the ring is variously tinted.

If urine is poured over cold acid, a precipitate may appear which is not albumin. This can happen when the urine is highly charged with urea, in which case crystalline nitrate of urea will separate out, or where urates are abundantly present, in which case the ring will consist of very fine crystals of uric acid, or acid urates. Both of these precipitates are dissipated by heat, and if the nitric acid is previously warmed, they cannot appear. It is better to make the test as just suggested than to use cold acid, and then try to warm a ring formed, as this would cause an admixture of the liquids sufficient to obscure a slight amount of albumin.

It is sometimes recommended to pour the urine in a test-tube, and by means of a pipette, or dropping-tube, allow the acid to flow under it. This is an excellent method of performing the test, but the acid should be slightly warm as before. If only a trace of albumin is present the ring will not appear immediately, but only after standing. It is well, therefore, in doubtful cases, to set the tube aside for twelve hours and then observe it. If a ring is now found it should be very gently and carefully warmed to determine its behavior toward heat, because on standing in the cold a ring of urates might appear.

When this test is applied to old, cloudy or alkaline

urine it should be preceded by this preliminary preparation :

Boil the urine with half its volume of 10 per cent potassium hydroxide solution and filter. This will usually give a bright, clear liquid, but if not add two drops of the "magnesia mixture" employed in qualitative analysis and described in the appendix, boil and filter again. The filtrate is now suitable for testing.

The action of the reagents is this : The strong alkali forms a bulky precipitate of the earthy phosphates present which usually settle and leave the supernatant liquid clear. The amount of alkali taken is sufficient to prevent the coagulation and precipitation of the albumin on boiling, while it serves also to expel ammonia which may be present. If the first filtrate is not perfectly clear the addition of the magnesia mixture accomplishes this by making a new precipitate of phosphates in traces which now leaves it bright.

With the clear filtrate the tests by addition of nitric acid may now be carried out. It must be remembered, however, that as the urine is now strongly alkaline, a relatively large volume of the strong nitric acid must be employed.

The text-books abound in minute descriptions concerning the best methods of conducting this comparatively simple test. The few sources of error which may mislead will now be pointed out. It is, of course, understood that these appear only in the search for small amounts of albumin, that is for amounts less than one-tenth of one per cent by weight. For greater quantities the reactions, even when not conducted with extreme care, are usually sharp.

When urine is poured over nitric acid or when the acid is introduced under the urine a *layer of some kind* always appears at the junction of the two liquids. The problem is to decide what this is. The peculiar appearance of a relatively large amount of coagulated albumin is so characteristic that any one who has ever seen it will recognize it again. But a faint cloud or haziness is, at the start, somewhat confusing. A colored layer or ring, which is very common, must not be mistaken for a precipitate or cloud. The normal urine coloring matters may produce a highly colored ring, and the bands with biliary colors are even deeper. But these color bands or zones are transparent which can be determined by holding the test-tube in the proper light.

Urine very highly charged with urea may give a crystalline precipitate of urea nitrate. This is a very unusual reaction, and the precipitate may be quite easily recognized through the form and size of the crystals, which are large flat plates readily seen by the naked eye or by a common magnifying lens. If a urine suspected to contain such an excess of urea be diluted with an equal volume of water before testing the crystals will not appear. Besides, they do not appear when the liquids are warm. The finely granular precipitate of acid urates or hydrated uric acid appears only in a cold liquid, therefore cannot be present to mislead if the test is conducted as directed.

If the special tests indicate the presence of unusually large quantities of urates the urine may be diluted with an equal volume of water before adding the nitric acid. It occasionally happens that a yellowish white cloud or band appears in this test which is not due to albumin or uric acid. Such a cloud may be caused by the presence in the urine of bodies taken into the system as remedies and which are excreted in but slightly changed form. Derivatives of turpentine and certain resinous bodies are specially liable to behave in this manner. After the use of copaiba balsam nitric acid throws out from the urine insoluble resin acids, which are not dissipated by heat. The precipitate formed by such acids dissolves readily in alcohol and can thus be readily distinguished from albumin which is not soluble. It must be remembered that while pure albumin precipitated by nitric acid is white, that thrown down from urine may be more or less colored from the presence of normal or abnormal coloring matters.

Attention must be called to a method of conducting the nitric acid test which is frequently employed, but which for small quantities of albumin is very untrustworthy. This method consists in mixing about equal volumes of strong acid and urine and boiling. This is open to the grave objection that by it the albumin sought may be decomposed and so lost from view. Nitric acid is a very strong oxidizing agent, and albumin a substance easily decomposed. Traces may therefore be lost even by a very short boiling, as may be readily determined by the student by a few experiments with weak albumin solutions.

Tanret's Mercuric-Potassium Iodide Test.

A solution of this compound precipitates albumin from acidified urine and is on the whole an extremely delicate reagent. Among the general albumin tests of Chapter IV. it was shown that many of these bodies are thrown out from their solutions in the form of complex basic compounds by addition of salts of certain heavy metals. Soluble salts of mercury, lead and copper give characteristic reactions.

Tanret prepared a well-known and popular test solution in the following manner:

Dissolve 33.12 Gm. of pure potassium iodide in about 200 Cc. of distilled water. Add 13.54 Gm. of powdered mercuric chloride and warm until, with sufficient stirring, the red precipitate of mercuric iodide disappears, leaving a clear, slightly yellowish solution. Dilute this with distilled water to about 800 Cc., and add 100 Cc. of pure, strong acetic acid. Allow to stand over night if not absolutely clear, and decant from any small precipitate which may have settled out. Dilute then to one liter with distilled water. This solution contains the two salts in the proportion of 4 KI to $HgCl_2$.

The test with the reagent so prepared is carried out as follows:

Filter the urine to make it perfectly clear, and add enough acetic acid to give it a good acid reaction. To about ten or fifteen Cc. in a test-tube add a very little of the reagent, a drop at a time, from a pipette or dropping-tube. In all not more than five drops should be added, as this is sufficient to give a strong precipitate if albumin is present. The precipitate is flocculent, and appears as a white cloud or streak, as the first drop of the heavy mercuric solution settles and mixes with the urine. As each following drop mingles with the urine the hazy cloud grows to a precipitate in case the urine contains more than a mere trace of albumin.

The delicacy of the reaction is remarkable. It is said that by it one part of albumin in one hundred thousand parts of urine may be detected. This, however, is probably excessive. One part in twenty-five thousand in a series of tests is nearer the average result. It has been claimed that where the solution is to be kept a long time it is best prepared without the addition of the acetic acid, as this is liable to produce slight decomposition in time. It is likely that the danger of this has been overestimated.

In any event, unless the urine is fresh and slightly acid the addition to it of acetic acid should not be neglected. The use of the acid is said by some writers to be unnecessary, but it has the advantage of disclosing the presence of any quantity of mucin which might interfere with the test.

If the acid throws out a cloud of mucin it should be filtered off and then the reagent added.

While this is an exceedingly valuable test certain precautions must be observed in its use. The mercuric solution is similar to one used as a test for alkaloids, and in fact precipitates many of these bodies. Quinine and other alkaloids given as remedies and excreted by the urine would therefore be shown by the test.

Alcohol dissolves these precipitates, however, but is without solvent action on that formed by albumin. Uric acid and urates give precipitates with the reagent if present in large amount, but such precipitates can be avoided by diluting the urine before testing, or if formed can be dissipated by slight heat.

Mistaking mucin for albumin can be avoided as shown above. Small amounts of peptones are precipitated by the reagent, but the coagulum disappears by application of heat.

The Ferrocyanide Test. A reagent of great delicacy is potassium ferrocyanide in presence of acetic acid. It shows not only serum albumin, but globulin and perhaps other proteids. It does not give a reaction with peptones.

The test is applied in this manner. The urine must be made as clear and bright as possible by filtration and then strongly acidulated with acetic acid. If a precipitate or cloudiness from mucin appears now filter again and to the filtrate add four or five drops of a *fresh clear* solution of the ferrocyanide. With even traces of albumin this gives a flocculent yellowish white precipitate.

One of the advantages claimed for this test is that it gives no reaction with the vegetable alkaloids and therefore can be used as a check upon some of the others, the mercuric-potassium iodide test for instance. The precipitate formed, although flocculent, is very fine and can be observed therefore only in clear solution.

A modified form of the test is the following: Mix five drops of the ferrocyanide solution with five Cc. of 30 per cent acetic acid. Pour this carefully over an equal volume of clear urine in a test-tube and allow to stand a short time. A white zone at the junction of the liquids shows the albumin.

The Picric Acid Test. Picric acid solutions, pure, or combined with citric, acetic or other acids, have long been used as reagents for the detection of traces of albumin in urine. In its simplest form the test liquid employed is a saturated aqueous solution of pure picric acid. It gives a very characteristic yellowish flocculent precipitate with even traces of albumin. Another solution frequently employed contains in one liter 10 Gm. of picric acid and 20 Gm. of citric acid. It must be made clear by filtration, if necessary, and is applied to clear urine in small quantity by means of a pipette, so as to show a cloudiness as the liquids mingle. The reagent is added gradually to the urine and in all not more than one-half the volume of the latter for a qualitative test.

The real practical value of this test is, in some quarters, still in dispute. It is certainly very delicate, but as it gives precipitates with peptones, alkaloids, urates, mucin, kreatinin, and perhaps other bodies, the first result observed is subject to revision. The precipitates formed with these substances and picric acid are dissipated by heat, but there is risk of getting the temperature too high, in which case other precipitates are liable to be formed, especially with the plain solution without the citric acid. A urine which yields a precipitate of earthy phosphates by warming will give it at the same temperature in the presence of picric acid. It seems to be true, however, that with citric acid added the interference from phosphates is eliminated, and there remains only mucin as a disturbing element. The danger here is not great, and it is likely that in all cases it can be avoided by adding the citric acid first, filtering if necessary, and then adding the picric acid.

Other Tests. A number of other tests are in use which show very minute traces of albumin. But they seem to possess no advantages over those enumerated above. One of these depends on the precipitation of albumin by phenol and acetic acid, in another picric and hydrochloric acids are used, in a third a strong solution of common salt and hydrochloric acid, and so on. But, practically, no one will find it necessary to go beyond the five tests given. Indeed, two are by most authorities generally thought sufficient, viz., the heat test and the nitric acid test. What cannot be shown by these reactions is so minute that for practical purposes it can be neglected usually.

THE AMOUNT OF ALBUMIN.

It is not alone sufficient that we are able to detect the presence of albumin in urine; we often need to know its amount to determine the practical value of a line of treatment pursued from day to day. To be of the greatest possible service, a method must be so easy of execution that approximately correct results may be obtained by it by the use of simple apparatus and in a short time. Several methods are known by which the amount of albumin in urine can be found. One of these, and the best, may be called the gravimetric method, as by it the albumin is precipitated, collected and weighed. In another, the albumin is precipitated and its volume measured, while in a third process the amount of albumin is estimated from the degree of turbidity caused by its precipitation in the urine.

Only the first and second methods will be described. The first is employed in exact investigations, and the second in clinical estimations.

The Gravimetric Method. If the qualitative test has shown only a small amount of albumin 100 Cc. of the urine should be measured out into a beaker for precipitation. If the qualitative test has given a strong indication 50 or 25 Cc. should be taken and diluted to 100. Enough dilute acetic acid is added to the urine to give it a *faint* acid reaction, after which it is brought up to a temperature of 80° or 90° C. on the water-bath, being stirred, meanwhile, frequently. From time to time the beaker is held up against the light, so that the operator may determine whether the coagulation is complete or not. A satisfactory coagulation is shown by the precipitation of the albumin in large flakes, leaving the surrounding liquid nearly clear. If this is not the case a little more acid should be added, *but very carefully*, and in all but three or four drops, unless the urine was strongly alkaline to begin with.

When the reaction seems to be complete on the water-bath the beaker is placed on gauze and the contents brought to boiling. The precipitate is then allowed to settle. Meanwhile, a small filter of well-washed filter paper is dried and weighed in a weighing tube. It is plaited and put in a funnel and then the albumin precipitate is collected on it. The precipitate is washed with hot distilled water until it gives no chlorine reaction to the wash water, then with absolute alcohol, and finally with ether. The funnel with contents is placed in an air oven and dried at 120° C. The filter is transferred to the weighing tube, and when cold is weighed. The increase in weight gives the albumin. Instead of collecting on paper, a much better plan is to collect on a Gooch funnel of asbestos, when it can be had, which simplifies the test, besides adding to its accuracy. This method consumes a good deal of time, but gives results which are near the truth when it is properly conducted. The best results are obtained when the weight of the precipitate does not exceed .3 Gm.

Volume Method. One of the simplest of these is the one proposed by *Esbach*. In this a special tube is used, called the *Esbach albuminometer*, and a special solution or reagent made by dissolving 10 Gm. of pure picric acid and 20 Gm. of pure citric acid in a liter of distilled water. The solution must be filtered if it is not perfectly clear, and is the same as the one used for the qualitative test. The principle involved in the employment of the method is this. The precipitate of albumin and picric acid settles in coherent manner and in a compact volume proportional to its weight provided certain definite amounts of the reagent and urine are taken. The albuminometer, or measuring tube used, resembles a test-tube of heavy glass about six inches long, and is graduated, empirically, to show how much urine and reagent to take and the amount of albumin obtained expressed in grams per liter, or tenths of one per cent. The annexed cut shows the tube and its markings.

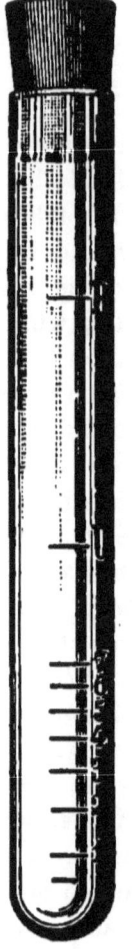

FIG. 41.

The test is carried out in this manner. Urine is poured in, to the mark U, and then the reagent, described above, to the mark R. The tube is closed with the thumb and tipped backward and forward eight or ten times until the liquids are thoroughly mixed. It is then closed with a rubber stopper and allowed to stand in a perpendicular position twenty-four hours. This will give the precipitate time to settle thoroughly, after which the amount can be read off on the scale. The results are accurate enough for clinical purposes and by practice can be made to agree moderately well with those found by the gravimetric method.

But to obtain this close agreement a number of precautions must be observed. The volume of the precipitate is in a marked degree variable with the temperature, and with the time given it for subsidence. The empirical graduation is based on the supposition that the test will be made at a temperature of 15° to 20° C, and that the reading will be made at the end of twenty-four hours. If the reading is delayed to two or three days the volume of the precipitate will be found much smaller. At the present time small centrifugal machines are rapidly coming into use to settle urine sediments. Some of these are operated by the Edison electric light current, and give the rotating tubes a high velocity. Where these machines are employed to settle the picric acid-albumin precipitate, the volume of the latter may be rendered abnormally small and the reading, therefore, prove erroneous.

The volume of the precipitate will depend here, not only on time and temperature, but also on velocity of rotation and the effect of this factor must be determined for each instrument before it can be accurately used.

In any case in applying this test the urine should not be highly concentrated. The best results are obtained with urine of low specific gravity and with the albumin not over .3 of one per cent. If a test shows an amount greatly in excess of this the urine should be diluted with a known proportion of water and tested again. On long standing in the cold a yellowish red precipitate of uric acid sometimes settles out. This need not mislead the student, as its color and general appearance are quite distinct from those of albumin.

The precipitates of albumin from urine by whatever means obtained are bulky and lead to the impression that the amount present is much larger than is actually the case. It was at one time customary to speak of 25, 30 and 50 per cent of albumin, these numbers representing the *apparent* volume of the precipitate in the test-tube. When these light precipitates are collected, properly dried and weighed a very different volume is obtained. A urine with one per cent of albumin contains an unusually large amount, and in any case seldom more than 10 or 15 grams of albumin occur in the day's urine. Between 1 gram and 5 grams are more common amounts, even in cases of acute albuminuria.

2. Serum Globulin.

This is an albuminous body resembling the serum albumin in many respects and often, perhaps generally, associated with it.

At the present time globulin in the urine has the same clinical significance as serum albumin. Although similar to each other in most points there are several characteristic differences which are taken advantage of as qualitative tests. The general relations of albumin and globulin were pointed out in a former chapter of this work and a number of reactions given for each. Not all of the tests for globulin which we find given in the books are suitable for use in the examination of urine, however, as we are here limited by the presence of other substances. Most of the reactions given above for albumin apply equally well to globulin and it is only within recent years that attempts have been made to detect one in the presence of the other. Among the methods applicable in the examination of urine the following may be given as most characteristic:

Qualitative Tests.

Dilution Test. Globulin is insoluble in water, but soluble in dilute salt solutions. Hence its solubility in urine. If the latter is diluted until the specific gravity is 1.002 or 1.003 the globulin may separate out. At any rate the addition of a few drops of dilute acetic acid will produce the desired result. A current of carbon dioxide passed into the diluted liquid for several hours accomplishes the same end.

The test may be modified in this manner. Filter the urine if it is not perfectly clear, and then pour it, *drop by drop*, into a tall, narrow beaker of distilled water. If globulin is present it is thrown out as a white cloud, which shows as the drops pass down through and mix with the lighter, clear water. The globulin may afterward be confirmed by adding a small amount of salt solution which will cause the precipitate to disappear.

Sulphate Test. Globulin may also be detected by reason of its insolubility in strong salt solutions. To this end treat the urine with enough ammonia water to give an alkaline reaction. This precipitates phosphates and sometimes other salts, which after a time are filtered off, leaving a clear liquid. To this add an equal volume of saturated solution of ammonium sulphate which, in presence of globulin, produces a white flocculent precipitate.

Magnesium sulphate is frequently used for the same purpose, and under some circumstances may possibly be preferable.

In this test there is some danger of confounding albumose with globulin, as the former is also precipitated by ammonium sulphate. But the danger of confusion is small if the conditions given are adhered to, *i. e.*, mix equal volumes of the clear filtered urine and saturated ammonium sulphate solution. For the precipitation of albumose higher concentration is necessary, as will be further explained below.

3. Albumose or Hemialbumose.

We have here a representative of an important class of proteid compounds which are derived from the albumins proper. It has already been explained that in the digestion of native albumins the albumoses appear as one of the stages, and are found therefore, among the products of peptic and pancreatic action in the body. It is likely that normally albumoses are always converted into peptones before leaving the alimentary tract.

The special significance of the bodies called albumose in the urine is by no means clear. While certainly a pathological appearance it has not yet been found possible to definitely connect it with any one disease. Its presence has been reported in the urine in several cases of osteomalacia but it appears by no means to be a constant accompaniment. Observers have called attention to its occurrence during the progress of several other diseases, but without being able to point out any definite relation.

Qualitative Tests.

The recognition of albumose is not a matter of difficulty, as it is distinguished from the other proteid compounds sometimes found in the urine by several well marked characteristics. It is not coagulated by heat or by the addition of acetic or warm nitric acid, and is very soluble in hot water. It is much less soluble in cold water, but the presence of small amounts of salts seems to increase its solubility here in marked degree.

In presence of albumin or globulin it can be found by the following process unless it is in very small amount. The urine is saturated with pure sodium chloride, and then enough acetic acid is added to give a strong acid reaction. The mixture is boiled and filtered hot. This treatment throws out both albumin and globulin, but does not precipitate any albumose present. The latter would therefore be found in the clear filtrate and sometimes in amount sufficient to precipitate as this cools. The filtrate should therefore be allowed to remain at rest until quite cool. If much albumose is present it will appear as a white cloud. Sometimes, however, it will be necessary to concentrate the filtrate before looking for this reaction, and this is done by evaporating slowly on the water-bath to half the volume. On now cooling salt will quickly settle out, while the albumose precipitates later in flocculent form.

Another test is this : Separate the albumin and globulin by boiling with a small amount of acetic acid without the salt. Filter while warm and concentrate the filtrate to a volume of one-third. Allow to cool thoroughly and add a large excess of *saturated* solution of ammonium sulphate. This gives a white flocculent precipitate of albumose, if present. The precipitate can be collected on a filter and washed with the saturated ammonium sulphate solution and then dissolved in a little distilled water, poured on the filter. This filtrate gives tests with picric acid, potassium ferrocyanide and acetic acid and other albumin reagents.

If the original urine shows no reactions for albumin or globulin the albumose tests can be applied directly after concentration. The method by precipitation by means of picric acid gives good results. The biuret test is also delicate and clear.

The Amount of Albumose.

This can be best found by collecting the ammonium sulphate precipitate on a filter and washing it thoroughly with more of the same saturated solution to remove urea and other nitrogen compounds. The neck of the funnel, with the moist precipitate, is then placed in the neck of a quarter-liter flask and distilled water poured on to dissolve the precipitate. From 50 to 100 Cc. will be amply sufficient to dissolve and wash the whole of the albumose into the flask below. The funnel is removed and distilled water added to bring the volume up to exactly 250 Cc. Of this volume 150 Cc. is transferred to a beaker, acidified with pure sulphuric acid and evaporated slowly down to a volume of 5 to 10 Cc. This is poured into a Kjeldahl digestion flask, the beaker being rinsed out with 10 Cc. of *pure*, strong sulphuric acid. Finally 10 Cc. more of the acid is added to the flask and then 10 grams of pure potassium sulphate. The flask is heated on the hot plate and the operation of determining nitrogen completed as described in a former chapter where the Kjeldahl method is explained.

The remaining 100 Cc. in the measuring flask is used for the determination of the amount of ammonium sulphate in the solution. This must be done in order to subtract the nitrogen corresponding from the result of the Kjeldahl distillation. The determination is best made by precipitating the sulphate with barium chloride in the usual manner.

4. Peptones.

As has been explained, peptones are proteid compounds formed from native albumins, globulins, fibrin, etc. in the digestive process. In this respect they are related closely to the albumoses, both differing from the other proteids in important respects. But peptones have further characteristic properties by which they are differentiated in turn from the albumoses.

Peptones may occur in the urine as a result of various abnormal conditions of the body; but their appearance does not depend, like that of albumin, on changes in the circulation or on pathological conditions of the kidney. Their clinical significance, therefore, is very different from that of albumin or globulin.

In cases of phosphorus poisoning the urine has frequently been found to contain peptones, but their presence can usually be connected with the disintegration of pus somewhere in the body. The peptone substances are, therefore, frequently, or perhaps generally, found in the urine in cases of purulent meningitis, purulent pleurisy, in the termination of pneumonia by resolution, and in general under circumstances in which products of suppuration can find their way into the circulation to be eliminated afterward by the kidneys.

Peptones have been reported in erysipelas, pulmonary tuberculosis, acute articular rheumatism, carcinoma of the gastro-intestinal canal, catarrhal jaundice, and in numerous other disorders. The condition in which the urine contains peptones as a result of the breaking up of purulent products is spoken of as *pyogenic peptonuria*, in contradistinction to that in which there is no indication of the existence of a suppurative process. It is claimed that in certain cancerous conditions of the stomach and intestine the peptones of digestion may find their way into the circulation.

Normally, peptones do not exist in the blood in more than traces, and during absorption from the healthy surfaces of the alimentary tract a change into albumin seems to take place. It is held by some writers that if taken up from an unhealthy surface (ulcerated) this conversion may not take place and peptone unchanged enters the circulation to disappear finally by way of the kidneys and through other channels.

It has been shown also that peptones are a normal constituent of the urine of women in the puerperal state and their occurrence has been pointed out under still other conditions not connected with a suppurative process. However, in the great majority of cases in which their existence is shown the connection with the latter is clear, and the detection of these substances becomes important as an aid to diagnosis.

Some of the reactions already referred to under "Chemical Physiology" can be applied to the urine, and additional ones are also available. The following tests may be applied:

QUALITATIVE TESTS.

Biuret Test. Peptones give this test as do other forms of albumin, but the final color is more nearly red than with the latter.

The direct treatment of the urine with strong alkali and the copper sulphate solution is seldom sufficient, it being usually necessary to subject it to a preliminary treatment to remove substances which interfere with the reaction. A large amount of peptone substance unmixed with albumin or globulin may give a very characteristic color, but in a highly colored urine this may be unsatisfactory and it is therefore safest to apply first a purifying process. The nature of the preliminary treatment will depend on the presence or absence of albumin or globulin. First, supposing these bodies absent we may proceed as in the following:

Hofmeister's Tests. In order to prove the presence of peptone in urine at least half a liter must be taken and treated with neutral acetate of lead in amount sufficient to produce a heavy flocculent precipitate, which is separated by filtration. An excess of the lead must not be used, but the solution is added carefully, a few drops at a time, giving the liquid meanwhile opportunity to settle so that a fresh precipitate can be seen after the addition of more of the reagent. By using proper care the right point can be found when addition of the acetate must cease.

Supposing now that we have a clear filtrate and that no albumin is present we next add to the filtrate a little hydrochloric acid and then a solution of phospho-tungstic acid in hydrochloric acid as long as a precipitate forms. If peptone is present it is contained in this precipitate which must be separated immediately by filtration, and washed on the filter with dilute sulphuric acid (5 per cent) until this passes through without color. The moist precipitate is then transferred to a small dish and mixed with a slight excess of barium carbonate or with enough crystalline barium hydroxide to give a slight alkaline reaction after thorough stirring. A little water is added and the whole is heated on the water-bath about ten minutes and filtered. The filtrate is then tested for peptone by the biuret test, or by the picric acid solution.

If albumin is present in the original urine it cannot be completely removed by lead acetate as just explained. The reaction with acetic acid and potassium ferrocyanide will usually show it in the filtrate from the lead and it must be removed as follows, this being accomplished by precipitating it with ferric oxide. Add to the 500 Cc. of filtrate a small amount of sodium acetate solution and then some ferric chloride, after which the urine is made neutral with sodium hydroxide and boiled. The iron should be completely precipitated, carrying with it the albumin. The solution is filtered, allowed to cool and tested for albumin.

If free from this it is ready for the peptone test, beginning with the addition of hydrochloric acid. If albumin is still present, add a little more iron or alkali, as experiment will decide, and boil again.

The phospho-tungstic acid solution is made up by various formulas, but as a reagent for urine the following method is recommended. A solution of pure sodium tungstate is made of about twenty per cent strength. To this is added either glacial phosphoric acid or syrupy phosphoric acid and the mixture boiled, the acid being added in sufficient amount to give a strong acid reaction. With glacial phosphoric acid the proportion should be about one part to four of the tungstate. The boiled mixture is allowed to cool thoroughly and to it is added about one-fifth its volume of strong hydrochloric acid. Allow the mixture to stand and pour off the clear liquid from any precipitate which may settle out. The complex phospho-tungstic acid is one of the few reagents which completely precipitate peptones and other albuminous bodies. In this case it is applied after the other albumins are thrown out by other means and serves to take the peptone away from coloring and equally objectionable substances.

Negative Tests. Peptones are not precipitated by heat or by the addition of hydrochloric, sulphuric, nitric or acetic acid. The reaction with potassium ferrocyanide and acetic acid, characteristic for the other albumins, is not given by the peptones.

They differ from the albumoses in not being thrown down by excess of ammonium sulphate, from which it follows that the test below can sometimes be applied for the detection of peptones in presence of albumin or albumose.

Precipitate 50 to 100 Cc. of urine, acidulated with a few drops of acetic acid, by boiling. Filter, concentrate the filtrate to a volume of 5 Cc. allow to cool and add 50 Cc. of cold saturated solution of ammonium sulphate and then some of the pure crystals so that the whole liquid is completely saturated. Then filter and to the filtrate apply the biuret test with a large excess of alkali and trace only of copper sulphate, or the phospho-tungstic acid test.

If the latter test is used the filtrate from the albumose should be diluted with two volumes of distilled water. The solution of phospho-tungstic acid gives a precipitate with normal urine, and also with the undiluted ammonium sulphate filtrate.

But reduced with water as directed no precipitate of the ordinary urinary constituents appears. A slight opalescence only may result, while in presence of even traces of peptones there is a marked precipitate. The reaction between phospho-tungstic acid and peptone solutions is one of extraordinary delicacy, so that traces show even after the above treatment.

5. Mucin.

In small amount mucin is probably present in all normal urines, in the case of women, coming especially from the vagina. In moderate amount it has, therefore, no pathological significance. If coming from the urinary tract in more than these traces it usually indicates an irritated condition or catarrh of the passages and then has clinical interest.

Urine containing mucin in large amount is turbid when passed; with a smaller amount it may be clear at first but on standing deposits a cloud which settles nearly to the bottom of the vessel and there floats in loose form, instead of compact as with other sediments. This cloud does not clear up by the addition of acetic acid or dilute nitric acid.

In clear urine containing mucin a flocculent, hazy precipitate is formed by the addition of acid. The test is best made by pouring some acetic acid into a test-tube and then carefully an equal volume of urine so as to mix the two liquids as little as possible. A mucin cloud appears in the urine layer above the zone of contact of the liquids. An albumin cloud makes its appearance lower, or at the zone. In testing for mucin in presence of albumin the main portion of the latter should be precipitated first by boiling and filtering. The mucin test can be applied to the cold filtrate by addition of acetic acid.

Mucin as well as albumin is precipitated from urine by the addition of an excess of strong alcohol, (three volumes to one). After some hours the precipitate may be collected on a filter and washed with alcohol. It is then washed with warm water which dissolves the mucin. This may be recognized in the aqueous solution by the addition of acetic acid.

Small amounts of mucin are so frequently mistaken for traces of albumin that attention must be paid to the methods of distinguishing between them. From what has been said it will be understood that the cloud which appears as a diffused haze in testing for albumin by an excess of acetic acid or a very small trace of nitric acid may be due to mucin and not to albumin as frequently assumed by mistake. A proper excess of nitric acid redissolves mucin, but not albumin in the cold.

Chapter XII.

THE TESTS FOR SUGAR.

ON the question of the occurrence of sugar in the urine a vast amount has been written. At one time, indeed until within quite recent years, it was generally assumed that normal urine contains no sugar or carbohydrate of any kind. But present methods of research seem to throw doubt on the truth of this view. It is not possible to separate small traces of sugar from a complex liquid like the urine so that the body separated may be recognized by its sensible properties. On the contrary we must depend on the results of certain reactions given by sugar solutions and in many instances by other organic bodies, and it is on the proper interpretation of these reactions that the authorities differ. Some of these reactions for traces will be explained below. In this place it suffices to say that the leading physiological chemists of the present time are nearly unanimous in holding that traces of the sugar known as dextrose exist normally in urine, in other words, that there may be such a condition as *physiological glycosuria* as distinguished from the well-known pathological condition characterized by the presence of relatively large amounts of sugar in the urine and named *diabetes mellitus.*

The amount of sugar believed to be normally present is very small and cannot be recognized by the first three or four tests given below. An amount of sugar in the urine sufficient to have clinical importance is readily recognized by many tests.

The characteristics of urine in true diabetes are these.

It has a specific gravity higher than normal, usually between 1.030 and 1.040, and *this with a greatly increased quantity*. A high specific gravity with small volume, it has been shown, need have no special clinical importance as such a condition can result from many causes outside of disease. Diabetic urine is usually light in color and prone to speedy decomposition by fermentation. The amount of sugar which can be present in advanced stages of *diabetes mellitus* may be very large. It is said that as much as 1,000 grams of dextrose has been passed with the urine in one day in extreme cases. But the amount usually coming under the observation of the practitioner is far below this, 10 to 100 grams being much more common amounts. In typical diabetes the *percentage* amount of the normal urine constituents is usually greatly diminished because of the great dilution with water, but the *actual* amount excreted in twenty-four hours may be increased.

It is well known that sugar may temporarily occur in the urine from a variety of causes. It has been found after the absorption of several poisons, and in cases of carbon monoxide poisoning; also in the course of certain diseases. The amounts present in these circumstances are usually small, and disappear with other symptoms of the disorder. The *continued* presence of considerable quantities of sugar is characteristic of only one disease, *i. e.*, the *diabetes mellitus*. This fact should be borne in mind in the practical examination of urine and tests should be repeated from time to time, unless the other clinical evidence is sufficient to immediately confirm the indication of the chemical test.

Qualitative Tests.

The tests for sugar in urine depend on several distinct general reactions. The most common of these reactions is that due to the oxygen absorbing power of alkaline dextrose solutions. The absorption of oxygen may give rise to a solution of characteristic color and odor as in the first one of the tests given below, or to certain precipitates formed by the abstraction of oxygen from metallic combinations in the test solutions employed. Some of these points have been already referred to in the first part of the book. The sugar tests which are most commonly employed in urine examination are the following:

Moore's Test. This depends on the reaction between grape sugar and strong alkali solutions. When a solution of sugar or diabetic urine is mixed, without heating, with a solution of sodium or potassium hydroxide, no change is at first apparent unless the amount of sugar present is large or the alkali very strong. But on application of heat, even with weak sugar solutions, a yellow color soon appears which grows darker, becoming yellowish brown, brown, and finally almost black, while an odor of caramel is quite apparent. The strong alkali-sugar solution absorbs atmospheric oxygen, giving rise to a number of products among which lactic acid, formic acid, pyrocatechin and others have been recognized. The brown color is due to other unknown decomposition products.

This is a good reaction for all but traces of sugar, as the intense dark brown color and strong odor are not given by other substances liable to be present in urine.

But traces of sugar cannot be recognized by this test with certainty, as the color of normal urine even is darkened to some extent by the action of alkalies.

Urine containing much mucin becomes perceptibly darker when heated with sodium, potassium or calcium hydroxide solutions.

The Trommer Test. This is one of the oldest and best known of the tests for the recognition of sugar in urine, and has been referred to before. It is performed by adding to the urine an equal volume of ten per cent solution of sodium or potassium hydroxide and then a *very few drops* (three or four to begin with) of dilute solution of copper sulphate.

Solutions of alkali and copper sulphate alone give a blue *precipitate* of copper hydroxide, but in presence of sugars and certain other bodies a *deep blue solution*, and not a precipitate is formed. Therefore, if the urine tested contains sugar the first indication is a more or less blue solution, stable for some time in the cold. On standing, however, the liquid turns greenish, and finally deposits a a yellow precipitate. This change takes place immediately on application of heat, the greenish colored precipitate turning yellow, and finally red by boiling. Copper suboxide precipitates, and this is the second and characteristic stage of the Trommer reaction.

Several substances can give the first stage, but dextrose is the only body liable to be present in the urine which can give a good indication in the second.

The test must, however, be used with certain precautions. Albumin, if present, must be coagulated and filtered off. The amount of copper sulphate used must be small, because if only a trace of sugar is present and much copper is used the latter will give a blue precipitate which does not redissolve, and which turns black on boiling, thus obscuring a sugar reaction which may be given at the same time.

In adding the copper sulphate it is best to pour into the test-tube containing the urine and alkali, first about three drops of a five per cent solution. If this appears to give a yellow color on boiling, which does not turn black more should be added, and this continued until a yellow or red precipitate is formed. A black precipitate on boiling shows that too much copper has been added, and that probably sugar is absent.

The active body in producing the reaction is copper hydroxide, but this must be in solution to act as a good oxidizing agent with sugar; and the test, therefore, becomes uncertain or unsatisfactory if so much copper is added that the hydroxide formed cannot be dissolved by the sugar which may be present. In doubtful cases it becomes necessary to make several trials before the right proportion between urine, alkali and copper solution is found. In the solution, on completion of the reaction, several oxidation products of sugar are found, among which are formic acid, oxalic acid, tartronic acid, etc. But the complete reaction is obscure. In order to avoid the indicated uncertainty of the Trommer test when used for small amounts of sugar the next one was proposed.

The Fehling Solution Test. *Fehling* suggested the use of a solution containing along with the copper sulphate and alkali a tartrate to dissolve the copper hydroxide formed by the first two. Many substances besides sugars, referred to in the last paragraph, have the power of dissolving copper hydroxide with a deep blue color. Among these may be mentioned tartaric acid and the tartrates, glycerol, mannitol and others of less value.

A solution prepared by mixing certain quantities of alkali, copper sulphate and either one of these bodies with water in definite proportions remains perfectly clear when boiled. But if a trace of dextrose (or several other sugars) is present the usual yellow precipitate forms. The preparation of Fehling solution proper is given in the appendix and its general use is explained in Chapter II. Here it suffices to indicate its special applications as a urine test. The great advantage which this solution has over the

Trommer test is found in the fact that it may always be used safely in excess. With only a trace of sugar there is no danger that the copper will precipitate as black hydrated oxide. In performing the test a few cubic centimeters of the Fehling solution (4 or 5) are poured into a test-tube, diluted with an equal volume of water and boiled. The solution must remain clear. Then the urine is poured in, at first about half a cubic centimeter, and the mixture boiled. If sugar is present in amount above one-tenth of one per cent it should show with the volume of urine taken. For smaller amounts of sugar more urine must be added, and the mixture boiled again.

When normal urine is heated with Fehling solution, a greenish flocculent precipitate usually makes its appearance. This has no significance as it is due to the phosphates normally present which come down when the reaction is made alkaline. Many urines produce a clear dark green solution when heated with the Fehling solution. This is a partial reduction reaction and like the other has no special importance as urines free from sugar give it. At other times urines free from sugar yield an almost colorless mixture when boiled with the Fehling solution. These peculiar reduction effects are due to the presence of uric acid, kreatin, kreatinin, pyrocatechin and several other substances and are generally characterized by discharge of the deep blue color of the solution without precipitation of the copper suboxide. Certain substances taken as remedies give rise to products in the urine which exert a similar action. Occasionally, however, the amount of uric acid is so large that the reduction is accompanied by actual precipitation of the copper as red oxide. This fact is of interest as it makes the test, at times, somewhat uncertain, but it is a very simple matter to determine whether or not a great excess of uric acid is present, as will be pointed out later. The liability to error in the Trommer test from these causes is less than in the Fehling test, but notwithstanding this the latter must still be regarded as the better test practically, because of its great convenience and the sharpness of the reaction with even traces of sugar. The ingredients of the Fehling test are best kept in separate bottles closed with rubber stoppers. A very convenient arrangement is explained in the following paragraph.

Two bottles, each holding about 200 Cc., are fitted with perforated rubber stoppers. Through the opening in each stopper the stem of a 2 Cc. pipette with very short tip is passed, and left in such a position that when the bottles are half filled the bulbs and stems to the mark will be covered with the liquid. One bottle contains the standard copper sulphate solution, the other the mixture of alkali and tartrate solution. The rubber stoppers should be covered with vaseline so that they will permit the pipette stems to slide easily in the perforations, and also close the bottles perfectly. When the stoppers are inserted the pipettes should stand full to the mark, ready for use.

On withdrawing the stoppers with forefinger closing the pipettes, exactly two Cc. of each liquid can be taken out without delay, and on mixing in a test-tube yield the Fehling solution, fresh and ready for use, directly, or after dilution with distilled water, as thought necessary. As the solutions are used the pipette stems are pushed farther through the stoppers so as to leave the marks always at the surface of the liquids. The solutions may be kept in this manner for years, and their use is not attended with any inconvenience. The open ends of the pipette stems should be kept closed with small rubber caps, or a bit of soft paraffine wax. The mixed Fehling liquid does not keep well unless prepared with certain unusual precautions, and therefore several other single solutions have been suggested, as described in the next paragraph.

Other Copper Solutions. The original Fehling solution has been modified in various ways. Most of these modifications consist in mere changes in the proportions of the ingredients dissolved. Two, however, may be considered as fundamentally different.

Loewe (1870) recommended a solution made by dissolving copper sulphate in water, adding solution of sodium hydroxide and then glycerol. For certain purposes copper hydroxide was found to possess advantages over the sulphate. The preparation of the Loewe solutions is described in the appendix. The claim was made by Loewe that the addition of glycerol prevents the spontaneous decomposition of the blue solution, which may, therefore, be kept mixed. While this is not absolutely correct it is true that the glycerol solutions keep much better than the mixed tartrate-alkali-copper solutions as usually made, and have therefore, found favor with some physicians.

Schmiedeberg (1886) described a solution containing in one liter 34.63 Gm. of crystallized copper sulphate, 16 Gm. of mannitol and 480 Cc. of sodium hydroxide solution of 1.145 Sp. Gr. This solution is easily prepared and has also the advantage of permanence.

Both the Loewe and the Schmiedeberg solutions have suffered slight alterations, without, however, being improved.

The second suggestion of Loewe, *i. e.*, to use copper hydroxide instead of sulphate, has not been generally followed, but it certainly has in some cases decided advantages. The final reaction in all these tests is the same as with the Trommer or Fehling test.

The Bismuth Test. *Boettger* found (1856) that in presence of alkali bismuth subnitrate is reduced to the metallic condition by the action of dextrose in hot solution.

As a urine test he recommended to make it strongly alkaline with sodium carbonate, and then add a very small amount, what can be held on the point of a penknife, of the pure bismuth subnitrate. On boiling the mixture the insoluble bismuth compound, which settles to the bottom, turns dark if sugar is present.

The test is at present carried out by adding to the urine in a test-tube an equal volume of 10 per cent solution of sodium or potassium hydroxide, and then the subnitrate. Boiling gives the reaction as before. In absence of sugar (or albumin) the bismuth compound remains white.

In performing this test only a very small amount of the subnitrate should be taken. This is absolutely necessary in the detection of traces of sugar. In this case the reduction is but slight, and not much black powder of bismuth or its oxide can be formed. If a great excess of the white subnitrate is taken it may be sufficient to completely obscure the reduction product. It is frequently well to use not more than four or five milligrams of the subnitrate.

The black precipitate formed was at one time supposed to be finely divided metallic bismuth. Later investigations seem to show that it consists essentially of lower oxides of bismuth. This test has certain advantages over the copper tests. It is easily made, and with materials everywhere obtainable in condition of sufficient purity. Furthermore, the reaction is not given with uric acid, which it will be remembered may act on the Fehling solution if excessive.

Albumin, however, interferes with the test, as it gives, also, a black precipitate when boiled with alkali and the bismuth subnitrate. In this case the albumin gives up sulphur and forms bismuth sulphide.

If albumin is present in a urine it should be coagulated and filtered out before trying the bismuth test. *Bruecke* recommends to coagulate by means of a solution of potassium bismuth iodide, the excess of bismuth serving to complete the sugar test. The reagent for this purpose is made by dissolving freshly precipitated bismuth subnitrate in a hot solution of potassium iodide by the aid of some hydrochloric acid. This is the solution previously recommended by *Fron* for the precipitation of alkaloids, and is made by dissolving 7 Gm. of potassium iodide in 20 Cc of water to which after heating 1.5 Gm. of the bismuth subnitrate and 1 Cc. of pure strong hydrochloric acid are added. The mixture must be kept hot until all is dissolved, resulting in an orange red solution.

This reagent precipitates albumin, but as it is rendered turbid by water the amount of acid necessary to prevent this for a given volume must be ascertained before it can be used with urine. This can be determined by adding a little of it (a few drops) to some water in a test-tube, and then dilute hydrochloric acid until the precipitate just disappears. The test proper is then made by taking the same quantity of urine and adding the same amount of acid and the reagent.

Albumin and other disturbing substances precipitate, and can be filtered off. The clear filtrate should not be made turbid by acid or the reagent. It is then made strongly alkaline with potassium or sodium hydroxide and then boiled. In presence of sugar a black precipitate is formed as before.

After adding the reagent it is necessary to wait several minutes for a possible precipitate to form and settle. The addition of alkali to the filtrate produces a bulky white precipitate of bismuth hydroxide which is readily reduced at the boiling temperature by sugar present. If only traces of sugar are present the boiling must be long continued to obtain the black precipitate. What was said above about the danger of obscuring this precipitate by the white bismuth compounds obtains also here.

When only a small amount of sugar is suspected it is best to allow the bismuth hydroxide precipitate to partially settle, and then to pour off the supernatant alkaline urine with a little of it. In this manner the amount of the bismuth compound which finally enters into the test is so small that it should all be reduced by even a trace of sugar, on subsequent boiling. When carefully performed this modification of the original Böttger test is a practically good one. It is not as sensitive as the Fehling test but shows traces of sugar of clinical importance.

The Phenylhydrazine Test. In this test a reaction discovered a few years ago has been applied by v. Jaksch to the examination of urine. Add to about 10 Cc. of urine .2 Gm. of phenylhydrazine chloride and a slightly greater amount of sodium acetate. Warm the mixture gently, and if solution does not take place add half the volume of water and heat half an hour on the water-bath. Then cool the test-tube by placing it in cold water and allow it to stand. If sugar is present a yellow precipitate settles out, which consists of minute needles generally arranged in rosettes, visible under the microscope. Albumin does not obscure this test, but if much is present it is best to coagulate it as well as possible by boiling and filter. The yellow precipitate is called phenylglucosazon.

For the detection of traces of sugar by this method it is necessary to use more urine and more of the reagents. 50 Cc. of urine with 2 Gm. of phenylhydrazine and 3 Gm. of sodium acetate may be taken.

Phenylglucosazon melts at 205° C. and a determination of the melting point may be made as a confirmatory test. For this purpose the supernatant liquid is poured off and the fine yellow crystals are washed with water by decantation. They are transferred to a small watch glass allowed to dry over sulphuric acid in a desiccator and are then ready for the test. Melting points are usually found by placing a small amount of the substance in question in a thin narrow tube, which is fastened to a thermometer by means of rubber bands. The substance in the bottom of the tube must be near the bulb of the thermometer. The bulb and bottom of tube are then immersed in a beaker of oil or sulphuric acid which is gradually heated until the substance begins to fuse. The temperature indicated by the thermometer is taken as the melting point. For best methods of working this test of finding the fusing point some standard manual of organic chemistry should be consulted.

On the whole, it must be said that this reaction is of very limited applicability in urine analysis. It has value only when the copper or bismuth methods are insufficient to decide concerning the presence or absence of sugar. In cases having real clinical importance such uncertainty is rare.

The A-naphthol Test. This depends on the reaction between α-naphthol and sugar in presence of sulphuric acid, and was discovered by *Molisch*. Take about a cubic centimeter of urine, previously diluted with five to ten volumes of water, and add to it two drops of a twenty per cent solution of α-naphthol in alcohol. Then add about half a cubic centimeter of strong sulphuric acid and agitate. A blue color indicates sugar. If the acid is carefully added so as to flow under the lighter liquid a blue zone is formed between them. By diluting largely with water and shaking, a violet precipitate is produced.

This method is exceedingly delicate, but unfortunately

is not characteristic, as many substances show the same result. The trace of sugar, or similar body, normally present, gives a marked reaction, hence the direction to largely dilute the urine before adding the reagent.

The α-naphthol may be replaced in this test by a twenty per cent alcoholic solution of thymol. The mixture becomes dark red and carmine red on dilution with water. It has been shown that these color changes depend on the formation of small amounts of furfurol by action of sulphuric acid on traces of carbohydrates and the subsequent combination of the furfurol with the α-naphthol or thymol. However, not only carbohydrates, but also albumins and many other substances yield furfurol in this manner and in normal urine some of these substances may be always present.

Molisch claims that the reaction found with highly diluted urine is a sugar reaction, that in condition of high dilution other bodies which may possibly be present cannot give this test. A color still shows when normal urine diluted 100 times is used, and on this behavior, partly, the claim that sugar is normally always present in urine is made. It will be seen from this that the test is too sensitive for ordinary clinical needs. But as a laboratory test it is valuable. By attentive study of the behavior of diluted normal and diabetic urines the chemist soon learns to recognize the deeper colors obtained with the latter and is therefore able to employ the test in the way of confirmation.

The Fermentation Test. When yeast is added to urine containing sugar and the mixture left in a moderately warm place the usual fermentation soon begins, which is shown by two principal changes. Carbon dioxide is given off, which may be collected and identified, and the mixture becomes lighter in specific gravity. When only traces of sugar are present the test by collection and identification of the carbon dioxide frequently fails because of the solubility of the gas in the liquid.

The variation in the specific gravity is an indication of greater value, as it can be readily observed with proper appliances. The test has practical value, however, only as a confirmation of some other one. If by the copper solutions, for instance, a strong indication is obtained which it is suspected may be due to an excess of uric acid, the reaction by fermentation may be resorted to because *only sugar* will respond to it. The test may be made by pouring 100 Cc. of the urine into each of two bottles or flasks. To one, half a cake of compressed yeast, crumbled, is added; the other is left pure. The bottle with the yeast is closed by means of a perforated stopper (to allow escape of gas), while the other is tightly corked. The two are left, side by side, in a warm place about twenty-four hours. At the end of this time a test of the specific gravity of the contents of both bottles is made. If sugar is present to the amount of one-half per cent the specific gravity of the yeast bottle should be perceptibly lower.

The test is frequently recommended as a quantitative one, as there is a fairly definite relation between amount of sugar and loss in density.

Other Sugar Reactions. Many other tests have been proposed for the detection of sugar in urine. A few of these will be referred to briefly in this place.

One of these, the picric acid test, is based on the fact that a urine containing sugar when mixed with solutions of potassium hydroxide and picric acid and boiled, turns a dark mahogany red from formation of picramic acid.

When urine is made strongly alkaline with potassium hydroxide and treated with a weak solution of diazobenzene sulphonic acid in water it turns reddish yellow, if sugar is present, and becomes afterward claret-red and finally dark red if much is in solution. The reaction is delicate, but is given by other bodies than sugar.

Another test depends on the reaction between sugar solutions and indigo-carmine in presence of alkali. The urine is made alkaline with sodium carbonate and treated with indigo-carmine until a deep blue is obtained on heating. If sugar is present on longer heating the color fades to yellow by reduction. The color returns by cooling and shaking with air.

These tests have given good results in the hands of those who have recommended them, but seem to possess no advantages over the copper and bismuth reactions.

The Amount of Sugar.

It is not always sufficient to be able to detect the presence of sugar in urine. A knowledge of the amount is frequently of the greatest importance. A number of methods have been proposed by which a quantitative determination can be made, some of them crude and of little practical value, while others give, when properly carried out, results which are accurate. The methods in general may be divided into four groups, depending on the

(1.) Reduction of solutions of heavy metals, and measurement of the amount of reduction.

(2.) Change of color produced in organic solutions, by action of sugar, the depth of final color being proportional to the amount of sugar.

(3.) Results of fermentation with measurement of change in specific gravity of the urine or measurement of evolved carbon dioxide.

(4.) Observation of rotary polarization of light.

Methods by Reduction of Metallic Solutions.

The reduction methods are illustrated in the use of the Fehling solution as a qualitative test and in the bismuth tests. The general principles involved in making a quantitative determination of sugar by aid of the Fehling solution have already been explained in the chapters on chemical physiology. When applied to the urine, however, the process requires certain modifications because of the fact that this secretion contains always a number of substances which interfere to some extent with the normal reduction and precipitation of the copper suboxide. The determination of dextrose in aqueous solution by the Fehling liquid is a problem of extreme simplicity, but in urine the case is somewhat different.

If we measure out 50 cubic centimeters of the mixed Fehling solution, heat it to boiling and then run in the saccharine urine from a burette it frequently happens that a greenish yellow *muddy* precipitate forms which does not turn bright red and which, instead of quickly settling to the bottom of the flask, remains suspended and makes it impossible to observe the disappearance of the blue color indicating the end of the reduction. This difficulty may be largely obviated by working with solutions of greater dilution, as explained in the following paragraph.

Determination by Fehling Solution. Prepare a Fehling solution as shown in the appendix and then accurately mix it with four volumes of distilled water. That is, to 100 cubic centimeters add 400 cubic centimeters of water, to 50 add 200, or to 25 add 100. In any event the dilution must be accurately made. One cubic centimeter of this liquid will oxidize almost exactly one milligram of dextrose as shown by a table in Chapter II., provided the sugar is in approximately one per cent solution. For all practical purposes of urine analysis the oxidizing power may be considered the same in a solution of one-half per cent strength, and only very slightly increased in still weaker solutions. Therefore, before beginning the test dilute the urine accurately with four or nine volumes of water. This can be done by making 50 Cc. up to 250 or to 500 Cc. and mixing well by shaking.

Now proceed with the analysis exactly as described in Chapter II. Measure out 50 Cc. of the dilute Fehling solution, pour it in a flask and heat to boiling on gauze. Fill a burette with the diluted urine and when the solution in the flask is actively boiling run in about 3 Cc. Boil two minutes, remove the lamp and wait half a minute to observe the color. If blue is still visible heat to boiling again and run in 3 Cc. more. After boiling two minutes as before wait a short time and observe the color near the surface of the liquid in the flask. If still blue repeat these operations until on waiting it is found that the blue has given place to a yellow. The urine should be so dilute that at least 10 Cc. must be run in to reduce all the copper hydroxide.

When the volume required is found to within 2 or 3 Cc. a second experiment must be made, the urine being added very gradually now, without interrupting the boiling longer than necessary, until the first of the limits between which the correct result must lie, as shown by the former test, is reached. From this point the addition of the urine is continued, with frequent pauses for observation of color until the reduction is complete. The volume of urine used contains 50 Mg. of sugar.

If the preliminary experiment shows that the urine is strong in sugar and that the reduction is easy, that is that the cuprous oxide separates and settles readily, the second test may advantageously be made with 50 Cc. of a stronger

Fehling solution. With many strong diabetic urines it is possible to use the undiluted copper solution with the oxidizing power of 4.75 milligrams of sugar to each Cc. The difficulties in this test have been very much overestimated; with a little practice any one can make a good sugar determination in urine. The important point is to find by a few simple preliminary tests the best conditions of dilution of Fehling solution and urine to give a precipitate which settles readily. With this information, and it can be acquired in a few minutes, the actual quantitative experiment can be easily made.

The Use of Pavy's Solution. To avoid some of the difficulties in the titration of diabetic urine by the Fehling solution, *Pavy* suggested a solution containing ammonia. If a solution of dextrose is run into a boiling copper solution containing ammonia in considerable quantity the copper is gradually reduced, giving finally a clear, colorless solution instead of a red precipitate. The end of the reduction is, therefore, indicated by disappearance of color alone. The preparation of the Pavy solution is given in the appendix. Its strength as there described is just one-tenth of that of the common Fehling liquid, that is, 100 Cc. oxidizes 50 milligrams of dextrose. The test is performed in a flask, as is the Fehling titration; but as the solution is easily changed by atmospheric oxidation, just as soon as it begins to hold some reduced copper, precautions should be taken to exclude the air during titration. This can be done by passing a slow current of illuminating gas or hydrogen through the flask during the test.

The titration is carried out as follows: Measure 100 Cc. of the ammoniacal copper solution into a flask holding about 300 Cc. Throw in some small pieces of pumice stone to prevent "bumping," and then heat to the boiling point on wire gauze. The sugar solution must be dilute, and should be contained in a burette with a delivery tip bent to one side and then down, so that the contents of the burette can be added slowly but continuously to the liquid in the flask without interrupting the ebullition. The operation should be carried out where there is a good circulation of the air to carry off the evolved ammonia fumes, and meanwhile a slow current of the hydrogen or coal gas should be led down into the flask to keep out the air. As the reduction is very slow the addition of the sugar solution must not be rapid. There is danger of adding too much until the operator becomes familiar with the method. If the precaution of passing in gas is neglected, which is usually the case, the results come out a little too low, because the air reoxidizes some of the ammoniacal cuprous solution, making it necessary to add more of the sugar to complete the reduction, that is, to completely discharge the color.

The method yields at best only approximate results, and working it subjects the analyst to the annoyance of ammoniacal fumes unless the apparatus is complicated by the addition of a delivery tube to carry the evolved ammonia through a window or into a fume chamber.

The reducing power of the copper in this solution depends to some extent on the amount of ammonia present, and from the fact that this is lost by ebullition during the performance of the test, irregularly and at different rates in different experiments, it follows that the results obtained cannot be perfectly uniform. Besides this the solution does not keep perfectly, its reducing power slowly undergoing change. However, the method has value and should be learned, because it can be rapidly worked and the results obtained are sufficiently accurate for clinical purposes.

The solution has been still further modified by substituting glycerol for the tartrate, giving what may be called the *Loewe-Pavy* solution. This solution is employed as is the Pavy liquid and has the same advantages and drawbacks. It is claimed for it, however, that it keeps somewhat better. Solutions containing ammonia cannot be used for qualitative testing.

Sugar Test by Solutions of Mercury. In Chapter II. it was explained that certain solutions of compounds of mercury can be used in sugar titration. Two such solutions are frequently used, viz., *Knapp's* solution, containing mercuric-potassium cyanide, and *Sachsse's* solution containing mercuric-potassium iodide. See the appendix for the preparation of both of these.

The solution of Knapp is frequently used in urine titration and is employed in the following manner: 10 Cc. of the solution, corresponding to 25 Mg. of dextrose is diluted with 25 Cc. of water in a flask and heated to boiling. The urine, which has been previously diluted accurately with from four to nine volumes of water, is run from a burette into the hot liquid until the whole of the mercury is precipitated, which can be recognized as follows: Allow the precipitate to settle and then by means of a glass rod place a drop of the yellowish supernatant liquid on a piece of white Swedish filter paper. Hold the paper then over an open hydrochloric acid bottle containing the fuming acid, and afterward over a beaker containing some strong hydrogen sulphide water. If the drop of transferred liquid contains even a trace of mercury this will be shown by the formation of a brown stain. In this case it will be necessary to add more of the sugar solution, and repeat the operations until the complete reduction and precipitation of the mercury compound is accomplished, as shown by negative result with the hydrogen sulphide test.

This method has been found to give very excellent results, but longer practice is necessary to give proficiency with it than with the other.

COLOR AND FERMENTATION METHODS.

The methods of quantitative sugar analysis depending on comparison of colors in sugar solutions acted on by picric acid and alkali or other reagent are neither very convenient nor accurate.

The fermentation test is sometimes applied quantitatively, but in those cases where it is the most accurate it is least necessary. With very weak sugar solutions it can only be used with the most careful regard to changes in temperature by the method referred to above. With strong diabetic urines accurate results are more readily reached, but here, by dilution, the copper solutions give the desired information more quickly and accurately.

When the saccharine urine is fermented as described and a change of specific gravity observed, the percentage of sugar is approximately given by multiplying each .001 lost by .23. For instance, if the urine before fermentation had a specific gravity of 1.032, and after fermentation, at the same temperature, a specific gravity of 1.016, we have $16 \times .23 = 3.68$ as the per cent of sugar present.

SUGAR DETERMINATION BY POLARIMETRY.

The construction and method of using the polarimeter or polariscope have been explained in Chapter II. It remains now to indicate how the instrument may be used in the examination of urine.

The direct examination of urine is not always possible because of its color and sometimes because of its slight turbidity. The best results are obtained with colorless and clear solutions. It is therefore sometimes necessary to prepare the urine by a preliminary treatment before it can be filled into the observation tubes. Diabetic urines light in color may frequently be used after simple filtration to render them perfectly clear, especially with the high class modern instruments of the half-shadow type with which a good illumination can be secured. If the urine is much colored, so that an observation cannot be made with the shortest tube—100 millimeters in length—which can be determined by a simple trial, resort must be had to precipitation to remove part of the color. Several precipitating agents are used for clarifying sugar solutions for the polariscope. The simplest of these is a solution of basic lead acetate which produces a voluminous precipitate that carries down much coloring matter. This is frequently used alone, but perhaps better combined with alum. When the basic acetate is added first and then some aluminum sulphate the mixed precipitate is flocculent and very effective in carrying down coloring matters. Use the basic acetate of lead described in the appendix and prepare a solution of aluminum sulphate of about equivalent strength, that is of such strength that one Cc. will precipitate the lead of one Cc. of the other.

Measure out 100 Cc. of the urine, add 5 Cc. of the lead solution and 2 or 3 Cc. of the alum solution, shake well, add water to bring the volume to 110 Cc. exactly, shake again and allow to stand 10 minutes. Then filter through a dry filter. The filtrate will be found much lighter in color than the original and probably suitable for use. If it is opalescent pour it through the precipitate on the filter when it will be found much brighter.

There is a slight loss of sugar in this operation as some is carried down by the precipitate. The clarified solution is then filled into the polarization tube and observed in the usual manner. The result obtained must be increased by one-tenth because of this dilution of the original urine. As the precipitate formed occupies an appreciable volume when dried, the clarified solution is correspondingly concentrated and the reading from this cause would be too high. For our purpose, however, we can assume that the gain in concentration is counterbalanced by the loss of sugar inclosed with the precipitate and neglect both sources of error.

If the urine contains albumin it must be separated by coagulation and filtered out, because it rotates the plane of polarized light to the left, and would therefore make the amount of sugar appear lower.

A given volume of urine is poured into a beaker and enough dilute acetic acid is added to give a faint reaction; it is then boiled, and after standing five minutes filtered. As all albuminous bodies, however, are not precipitated by simple coagulation with acetic acid, it has been recommended to add to 100 Cc. of the urine, 10 Cc. of the strongly acid solution of phospho-tungstic acid, already described, and filter after ten minutes. The dilution must be allowed for in the final calculation. Some coloring matters are also removed by this treatment. The urine may contain other active substances, but in amount so small that their effect may be neglected entirely.

Calculation of Result. For sodium light the formula

$$[a] = \frac{100a}{lc}.$$

is used with the factor $[a] = 53°$

Hence we have $c = \dfrac{100a}{l.53°}$

That is, the number of grams of diabetic sugar in 100 cubic centimeters of the solution polarized is equal to the product of the observed angle of rotation multiplied by

100 and divided by the product of the length of the observation tube in decimeters multiplied by the specific rotation, 53°

If in a given case we find a rotation of 10° 36′, with a tube two decimeters long, our formula becomes

$$c = \frac{100 \times 10.6}{2 \times 53} = 10$$

that is, the concentration, c, is 10 grams per 100 Cc.

With a decimeter tube each degree of rotation corresponds to a concentration of 1.8868. With the usual two decimeter tube each degree indicates 0.9434 Gm. in each 100 Cc.

The specific rotation of dextrose as obtained from urine appears to be a little higher than is that of the product made from starch.

Other Sugars in Urine.

Pathologically, traces or even larger quantities of several other saccharine bodies are occasionally found in urine. Among these we have first:

Lævulose, or Fruit Sugar. This is found along with dextrose in some cases of diabetes, but does not appear to occur alone.

While the recognition of lævulose in the pure state or in simple aqueous solution is a matter presenting no difficulty, the certain detection of this body as it occurs in urine is by no means as readily effected. This sugar gives the reduction and fermentation tests as described under dextrose, and therefore, cannot be distinguished by these methods. Lævulose, however, rotates the plane of polarized light to the left, and this property is sometimes of service in aiding the recognition. If the rotation is strongly to the left, the presence of lævulose in quantity may be inferred, assuming that albumins are absent. If the quantity of sugar, calculated as dextrose, determined by polarization is much below that found by the copper reduction method the indication is that lævulose is present with the dextrose. An exact measurement of the amounts of the two sugars when mixed in the urine is not possible with present means.

Lactose, or Milk Sugar, is occasionally found in the urine of nursing women. Its certain detection when in small amount presents even greater difficulties than is the case with lævulose. As its rotation is right-handed the polariscopic test is of little value.

Milk sugar is more strongly acted on by Fehling solution than is dextrose. While 1 Cc. of the copper solution oxidizes 4.75 Mg. of dextrose, it oxidizes 6.76 Mg. of milk sugar.

When a solution of milk sugar is boiled with dilute hydrochloric acid it yields dextrose and galactose, the latter resembling dextrose in its behavior with the copper solution. The specific rotation, $[\alpha]_D$, of dextrose is $52.7°$, of lactose, $52.5°$, while that of galactose is $81.3°$. The specific rotation of a mixture of equal parts of dextrose and galactose has been found by experiment to be $67.5°$, which agrees closely with the mean of $52.7°$ and $81.3°$. If, therefore, the rotation of urine is increased after heating with acid and neutralizing, and its copper reducing power diminished, we have data suggesting the presence of milk sugar. Experiments to show these points with certainty must be very carefully conducted, consuming no little time in manipulation. They are, therefore, of little value from a clinical standpoint.

Inosite, or Muscle Sugar, has been found in urine, in

diabetes, and also with albumin. There is no simple method by which it may be separated in the small quantity in which it occurs in urine.

Dextrine and a body termed *animal gum* have been reported as occurring in some cases of diabetes, but as their clinical significance is not clear nothing more need be said about them here.

Acetone.

This is a substance which frequently is found in urine in small amounts. Indeed, it may be true, as has been asserted, that it is normally always present in traces. This *physiological acetonuria* has no clinical significance. Under some circumstances, however, it may be found in larger quantity, sometimes in amount sufficient to be detected by the odor alone, which fact first called attention to it. At one time it was supposed to be related to the sugar found in urine, but it is now established that it more generally accompanies albumin and is frequently observed in many febrile conditions.

Acetone in urine is believed to be a decomposition product of albumins. It has been shown that in health, even, it can be much increased by a diet rich in nitrogenous materials.

But, occurring as it does in fevers and in advanced stages of diabetes mellitus, a certain interest attaches to its detection, and numerous methods have been proposed by which it may be identified in small amount. Those which depend on its direct recognition in the urine are mostly uncertain. It is always safer to distill the liquid and apply the test to a portion of the distillate. Half a liter, or more, of the urine is poured in a retort attached to a Liebig's condenser, and after addition of a little phosphoric acid is subjected to distillation. One hundred cubic centimeters of distillate will be enough. A portion of this can be taken for each test as follows:

Legal's Test. Add to 25 Cc. of the fluid a small amount of a fresh solution of sodium nitroprusside, and a few drops of a fifty per cent potassium hydroxide solution. If a ruby red color appears which slowly gives place to yellow, and if the addition of acetic acid changes this to purple, or violet red, the presence of acetone is indicated.

Lieben's Test. This depends on the production of iodoform, and is carried out in this manner. To about 5 Cc. of the distillate add a few drops of a solution of iodine in potassium iodide (the "compound solution of iodine," Lugol's solution), and then a small amount of potassium hydroxide, to marked alkaline reaction. If acetone is present a yellowish white precipitate soon appears, which, on standing, becomes crystalline and more deeply colored. The test is said to be sharper and more characteristic if ammonia is used instead of the fixed alkali. The liquid is first made strongly alkaline with ammonia, and then the iodine solution is added until the brownish precipitate formed at first dissolves very slowly. In a short time the yellowish iodoform precipitate makes its appearance. A rough quantitative measure of the amount of acetone present is given by noting the smallest volume of the distillate with which a distinct iodoform reaction can be seen. It is said that .0001 Mg. in one Cc. can be detected. .5 Mg. in 10 Cc. can be recognized by the nitroprusside reaction.

Kreatinin gives a ruby red color as does acetone when the nitroprusside reaction is directly applied to urine, but after adding acetic acid a green or blue color results.

Aceto-acetic or Diacetic Acid.

This compound is very frequently found associated with acetone in the urine of fevers and in diabetes mellitus. While acetone may occur in very small amount normally it is believed that aceto-acetic acid is always pathological. In the past few years much has been written on the subject of this substance and its clinical significance. It appears from the discussion that its presence in diabetes is of especial importance and that any increase in its amount should be carefully followed by analytical tests. What is known as the coma of diabetes is closely associated, according to eminent authority, with the presence of aceto-acetic acid in the blood.

Just how this body is produced in the blood from which it passes into the urine cannot be explained satisfactorily at present. Several theories have been offered to account for the phenomenon, but they are scarcely definite enough to be presented in an elementary work on practical tests like the present.

Urine containing aceto-acetic acid always contains acetone. The latter is probably derived from the former, and both may be derived from a still more complex substance, β-oxybutyric acid. It has lately been claimed that many of the reactions supposed to be due to aceto-acetic acid are in reality due to this substance. For our purpose, however, it will be sufficient to follow the important tests by which the aceto-acetic acid may be recognized.

The simpler acetone test should be made first because the other appears to be present only with this, and because further, the aceto-acetic acid gives these tests as well as does acetone.

Ferric Chloride Test. Our main test for aceto-acetic acid depends on a reaction with ferric chloride with which it strikes a red color. Normally, there is nothing in urine which gives the same reaction, so that if on the addition of a few drops of solution of ferric chloride to fresh urine a wine red color results the presence of aceto-acetic acid may be inferred. At the present time, however, many coal tar products are given as remedies which oxidize to compounds that, on elimination with the urine, give a red or purple color with ferric chloride when added. To detect the aceto-acetic acid with certainty under these conditions it is necessary to proceed with greater care. To this end add to the urine, which should be fresh, a few drops of ferric chloride or enough to precipitate the phosphates present. Filter and add a little more of the chloride. A red color indicates the acid. Divide the liquid into two portions; boil one and allow the other to stand a day or more. In the boiled portion the color due to aceto-acetic acid should disappear within a few minutes, while in the other it should remain about twenty-four hours.

Acidulate another portion of the urine with dilute sulphuric acid and extract it with ether which takes up aceto-acetic acid. Remove the ethereal layer and shake it with an aqueous solution of ferric chloride. The red color should appear as before and disappear on boiling, which behavior distinguishes the acid from other substances likely to be present.

Chapter XIII.

THE COLORING MATTERS IN URINE. BILIARY ACIDS.

Normal Coloring Matters.

ALTHOUGH many investigations have been carried out on the subject of the normal urinary pigments we are yet unable to give a very definite account concerning them. This is partly due to the fact that the coloring substances exist in the urine in minute traces only, which makes their separation and recognition exceedingly difficult, and partly to another fact that some of them are easily altered or destroyed by the action of the reagents employed in their investigation. By proceeding according to different methods, physiologists have obtained very different results indicating the existence of several colors, or at any rate modifications of colors. It is generally admitted, however, that at least two distinct coloring matters exist in the urine, and the others may be classed as derived from these by oxidation processes. One of these colors is known as *urobilin* and the other as *indican*.

Among the products related to or derived from urobilin the following may be mentioned as described by different writers. Urohæmatin, *Harley*, urophain, *Heller*, hydrobilirubin, *Maly*, urochrome, *Thudichum*. Among the colors related to indican, uroxanthin, *Heller*, may be mentioned.

Urobilin. This has been obtained as a reddish brown amorphous substance, but probably not in absolutely pure condition. It is slightly soluble in water, readily soluble in alcohol and chloroform. The *neutral* alcohol solutions are characterized by a marked greenish fluorescence which is an important means of recognition. The *acid* alcohol solutions are reddish in color, the shade varying with the concentration.

If present in more than minute traces in urine it gives characteristic absorption bands in the spectrum which have been referred to before. In acid urine the center of the dark band is near the Fraunhofer line F ; in alkaline urine the center is about midway between b and F.

Urobilin is generally much increased in fevers and in some diseases of the liver and heart. Any cause tending to break up the red corpuscles, increases urobilin. It is not always present in sufficient quantity in normal urine to be easily recognized. If the quantity is abnormally large the following test will show it :

Add ammonia water to strong alkaline reaction and filter if necessary. Then add a few drops of solution of zinc chloride, but not enough to give a permanent precipitate. In this way a zinc salt is formed, which shows a peculiar greenish fluorescence.

Ammonia generally causes a precipitate of phosphates, hence the direction to filter. If the characteristic fluorescence fails to appear the following modification may be tried, which is sufficient to give the reaction with most urines.

Precipitate 200 Cc. of urine with basic lead acetate, collect the precipitate on a filter, wash it with water and dry it. Then wash it with alcohol.

Finally, digest with alcohol containing a little sulphuric acid, and filter. The filtrate is usually fluorescent. Make it strongly alkaline with ammonia, and add solution of zinc chloride. This will give the fluorescence referred to above if but little is added, while if an excess of the zinc chloride is added, a reddish precipitate falls.

Urophain. This is the name given by Heller to a substance identical with, or similar to, urobilin. Heller gives this test: Take a few Cc. of strong sulphuric acid in a conical glass and pour on it, drop by drop, about twice as much urine. As the two mix, a deep garnet red is produced.

This reaction is not, however, characteristic, as several other matters may give it.

Urohæmatin is the name given by Harley to a coloring matter similar to the above. He applies this test: Dilute or concentrate the urine so that it is equivalent to 1,800 Cc. for the twenty-four hours. Take a few Cc. in a test-tube or wine glass, and add one-fourth of its volume of strong nitric acid. No change of color can be observed if the urohæmatin is present in normal amount. If more than this is present various shades from pink to red may be produced. The test should be made with cold urine, as with increased temperature darker colors result.

Indican and its Reactions. Although a normal constituent of urine indican is found greatly increased during the progress of certain diseases and becomes therefore a substance of clinical importance. It is formed along with other complex compounds in the oxidation of indol in presence of sulphuric acid. Indol is one of the common products of putrefaction, a change brought about in albuminous bodies, usually by bacterial agency. Such changes may take place in the alimentary canal, and the indol formed becomes oxidized to indoxylsulphuric acid or indican, and appears as such in the urine. The sulphuric acid necessary for the production of this body is present in combination in the system.

If much indican is found it suggests that abnormal

putrefaction is taking place somewhere in the body. In diseases accompanied by the formation of putrid secretions indican usually appears in increased amount, and hence the inference derived from its ready detection. It is found in increased amount in cancer of the stomach or liver, in peritonitis, in some stages of pleurisy, in intestinal invagination (whereby the normal passage of albuminous and other food products is hindered, thus making putrefaction possible) and in other diseases.

Indican is found in normal urines in very small amount only. It may, under favorable circumstances, be detected as here given: Take about four Cc. of pure hydrochloric acid in a test-tube and add about half as much urine, shaking well. A blue or violet color shows indican.

A more generally applicable method is this: To 10 Cc. of urine and the same volume of strong pure hydrochloric acid, add 2 or 3 Cc. of chloroform. Then add, drop by drop, solution of sodium hypochlorite, shaking after each addition. The hypochlorite acts as an oxidizing agent, liberating the coloring matter, which is then taken up by the chloroform.

The oxidation must not be carried too far; that is, too much hypochlorite must not be added, as it would then destroy the color as fast as formed.

Albumin must be separated by coagulation before applying either of these tests, as it develops a blue color with hydrochloric acid. The amount of indican normally present in urine is said to vary between 5 and 20 milligrams daily. The chloroform layer in the bottom of the test-tube in the above test shows roughly by the depth of color developed the amount of indican present. It is necessary to use good hypochlorite for this test as with a weak solution the oxidation may fail to take place.

Abnormal Coloring Matters.

In disease several other coloring matters may appear in urine, the most important of which are those of the bile and blood.

As abnormal colors must be classed, also, many products taken into the stomach with the food or as remedies and which appear directly in the urine or give rise to marked coloration on the addition of reagents.

BILIARY COLORING MATTERS.

These are found in the urine in jaundice and may be traced to the stoppage of the bile ducts of the liver as in common jaundice and to other causes having no connection with a disorder of the liver. The appearance of these coloring matters in the urine is therefore a symptom of different diseases, although perhaps most commonly associated with an abnormality in the flow of the bile. Jaundice may sometimes be traced to a disintegration of the red corpuscles in the blood and consequent liberation of derived coloring matters.

Biliary urine has generally a characteristic greenish yellow color sometimes tinged with brown. The froth from such urine is readily recognized by its yellow color, which is often a sufficient test in itself. Among the chemical tests the following are the best known.

Gmelin's Test. This has been referred to before and depends on the oxidation of bilirubin, the pigment commonly present in fresh jaundice urine, by nitrous acid. Pour in a test-tube about 5 Cc. of the urine under examination and by means of a pipette introduce below it an equal volume of strong nitric acid mixed with nitrous. This should be carefully done so as to avoid mixing the liquids much. At the junction of the two liquids, if bile is present, several colored rings appear of which the green due to biliverdin is most characteristic. Bands of blue, violet, red and yellow may appear above the green, but this, next to the acid where the oxidation action is strongest, is essential. It must be remembered that nitric acid gives the other colors at times with urine free from bile, but green is characteristic of the latter.

Fleischl modified this test by mixing the urine with a strong solution of sodium nitrate and then adding strong sulphuric acid carefully. This settles below the urine and decomposes the nitrate at the point of contact liberating the necessary nitric and nitrous acids for the oxidation as before. This method is a very good one.

In another modification, urine is dropped on a plaster of Paris disc and then a few drops of the oxidizing mixture of nitric and nitrous acids is placed in its center. The same play of colors appears as before.

Trousseau's Test. Add to some urine in a test-tube a few drops of tincture of iodine, allowing the iodine to float on the urine. If bile pigments are present a green color is produced when the iodine touches the urine, and persists some hours. Care must be taken to avoid using an excess of the iodine if the fluids are allowed to mix. In this case with the proper amount of the tincture the whole urine appears green.

Heller's Test. Take 5 or 6 Cc. of pure strong hydrochloric acid in a conical glass and add enough of the urine to give it a faint color on mixing. Now add pure nitric acid by means of a pipette so as to bring the latter under the mixture of hydrochloric acid and urine. The colored rings appear as in the Gmelin test and on shaking can be followed through the liquid.

The Detection of Traces. To 100 Cc. of the urine add 10 Cc. of pure chloroform and shake gently until the latter is colored. By means of a pipette withdraw a small part of the chloroform and mix it in a test-tube with 10 Cc. of strong pure hydrochloric acid. Add nitric acid as in the other tests and shake. With bile present the oxidation colors appear slowly in the chloroform, the green being the deciding tint.

BLOOD COLORING MATTERS.

As these appear in the urine they may be derived from different sources. We may have, first, color due to the presence of blood corpuscles themselves sometimes in nearly fresh condition. There may be enough blood present to impart to the urine a marked red color and it may be derived from the kidney, bladder, urethra or other part of the urinary tract. In blood from a fresh lesion the corpuscles usually appear in clearer outline than is the case when they have remained long in contact with the urine.

The presence of blood may be detected by several methods. The corpuscles are often easily recognized by the microscope in the sediment deposited when the urine is allowed to stand, as will be explained in a following chapter. Then we can make use of the spectroscope by which means the characteristic absorption bands of oxyhæmoglobin are detected, as was shown in the earlier chapter on the blood. If urine containing blood is treated with a few drops of ammonium sulphide and very gently warmed the spectrum of reduced hæmoglobin is given.

Sometimes the coloring matters alone without the corpuscles can be found. This is the case when the latter become disintegrated, the more stable and soluble hæmoglobin passing into solution while the stroma disappears by decomposition. The condition in which blood itself is present, and can be recognized by the microscope, is known as *hæmaturia*, while the condition characterized by the presence of the coloring substance only is called *hæmoglobinuria*.

In urine, hæmoglobin frequently undergoes two decompositions. It may become converted into *methæmoglobin*, as was explained in Chapter V., or it may suffer a complete modification, breaking up into hæmatin and a body resembling globulin. Hæmatin is best recognized by spectroscopic examination, as it gives a spectrum different from hæmoglobin. This modified product is said to occur in urine in cases of poisoning by hydrogen arsenide.

The following are the best chemical tests for the recognition of these bodies :

Heller's Test. Treat the urine with solution of sodium or potassium hydroxide, and heat to boiling. This produces a precipitate of the earthy phosphates which in subsiding carry down coloring matters. If a precipitate does not separate readily it may be hastened by adding two or three drops of magnesia mixture. Hæmoglobin, when present, is decomposed by this treatment with separation of hæmatin, which in turn settles down with the phosphates, imparting a red color to the precipitate.

Struve's Test. Make the urine slightly alkaline with sodium hydroxide solution, and then add enough solution of tannic acid in acetic acid to change the reaction. If hæmoglobin is present a dark brown precipitate of hæmatin tannate settles out. The test is a good one, and easily performed. This precipitate can be used for the production of Teichmann's hæmin crystals by moistening with salt and hydrochloric acid by the method described in Chapter V.

Almen's Guaiacum Test. In a test-tube mix equal volumes of fresh tincture of guaiacum and ozonized turpentine. Two or three Cc. of each will suffice. The mixture, if made of proper materials must not show a green or blue color after thorough shaking. Now add a few Cc. of the urine to be tested, a drop at a time, and agitate after each addition. If hæmoglobin is present it causes the oxidizing material of the ozonized turpentine (probably hydrogen peroxide) to act on the precipitated guaiacum resin, imparting to it first a greenish, and finally a blue color. Old and alkaline urine must be made faintly acid before performing the test. Pus in the urine gives a somewhat similar reaction, and a few other bodies, very seldom present, interfere. The test is very delicate, and if it gives a negative result it is safe to conclude that blood is absent.

Vegetable and Other Colors.

It has long been known that many peculiar coloring matters enter the urine from substances taken as remedies and sometimes as food. A few of the more common of these colors will be mentioned here.

Chrysophanic Acid. This complex organic acid is found in the root of several kinds of rhubarb, in senna leaves, in certain lichens and elsewhere. After the administration of any of these substances the urine becomes more highly colored, being a brighter yellow if acid and yellowish red when made alkaline. When phosphates are precipitated by addition of alkali they appear red in presence of chrysophanic acid, as they do with blood. But the latter can be easily distinguished by the other tests already given.

Santonin. This crystalline principle is found in the unexpanded flowers of Levant wormseed, and when administered as a remedy produces a characteristic change in the color of the urine. The color becomes a deep yellow which turns red with alkalies, as in the case of chrysophanic acid. If the colored alkaline urine is shaken with amyl alcohol the coloring matter from the santonin leaves the urine and passes into the alcohol, but the color from chrysophanic acid is only very slightly soluble in amyl alcohol and remains with the urine when the same treatment is applied.

Salicylic Acid. The urine of persons taking this substance has usually a grayish smoky tinge which becomes blue on addition of solution of ferric chloride if more than traces are present.

Salicylic acid is excreted in the free state or as a salicylate of sodium or potassium mainly; a small portion seems to pass into other compounds. But as the iron reaction is very delicate minute amounts of the free or combined acid can be found.

Enough ferric chloride must be added to be in excess of what would combine with the phosphates present, otherwise a sharp reaction may not be secured.

Phenols. Several phenol bodies as carbolic acid, hydroquinol, resorcinol, pyrocatechol and others sometimes find their way into the urine, to which they impart a dark color on standing exposed to the air. This change of color is said to be due to the formation of oxidation products of hydroquinol. From urine darkened in this manner phenols have been recovered by making acid with sulphuric acid and then distilling with steam.

Alkapton. In some cases described in the literature in the last few years the urine has had a brownish tint turning darker on exposure to the air. The substance giving rise to this color was called *alkapton*, but has been shown to be a mixture, probably, of homogentisinic acid with one or more other complex aromatic products. The color is very marked in presence of alkali and can become almost black, absorbing oxygen in this condition rapidly. The clinical significance of this substance is not well understood as yet.

Other Colors. Blueberries, carrots and several other common vegetable foods give deep color to the urine. It is not always possible to recognize the coloring substances in these cases. Such urine usually becomes yellow with acids and reddish with alkalies. It is occasionally possible to identify the color by means of the spectroscope, as the absorption spectra of some of these products have been studied.

The Detection of the Bile Acids.

It sometimes happens that the physician desires information regarding the presence of the biliary acids as well as the pigments in the urine. This information, however, is not easily secured because there is no simple test which can be applied directly to the urine which will give a certain indication of the presence of these acids. They must first be separated from the large amount of other substances present, which can be done in this way (*Neukomm*):

Evaporate 300 to 500 Cc. of urine nearly to dryness; extract with ordinary alcohol, evaporate this solution, and extract the residue with absolute alcohol.

Evaporate this and take up the residue with water. Precipitate the solution by lead acetate, *avoiding excess;* allow to settle, wash with water on a filter, and dry by pressing between bibulous paper. This leaves an impure lead salt of the acids. Extract it with hot alcohol, and filter; add sodium carbonate to the filtrate, evaporate to dryness and extract the sodium salt, thus formed, with absolute alcohol. Evaporate again, add some water and apply the *Pettenkofer* test, as follows:

To the solution add one or two drops of a 20 per cent cane sugar solution, and then some strong sulphuric acid, slowly to avoid heating.

It is best to immerse the test-tube in water to keep the temperature below 60° C. As the acid mixes with the liquid a violet or purple color is produced. It has been shown that this, like the naphthol test for dextrose is a furfurol reaction, the furfurol formed from the mixed sugar and acid combining with the acids of the bile. It has even been proposed to use a dilute solution (one-tenth per cent) of furfurol instead of the sugar in the test.

Kuelz recommends to evaporate the solution on a water-bath to dryness, to moisten the residue with a drop of dilute sugar solution, and then with a drop of the strong acid. The color appears almost immediately, but can be sharpened by heating the evaporating dish a few seconds on the water-bath.

Applying either of these tests directly to urine is unsafe, as the coloring and other matters present would interfere very much with the reaction.

Chapter XIV.

DETERMINATION OF URIC ACID.

URIC acid occurs normally in urine combined with sodium, potassium, magnesium, or ammonium. The absolute amount excreted daily is small but quite variable, depending on many conditions not well understood. In health the amount passed daily seems to vary between 0.2 Gm. and 1 Gm. These limits may not be correct, however, as many of the older determinations were made by inaccurate methods.

Regarding the clinical significance of variations in the amounts of uric acid passed our knowledge is still very defective. It is generally held that there is a considerable increase in the excreted uric acid in fevers and in diseases characterized by diminished respiration and consequently imperfect oxidation. In leucæmia there is a pronounced and characteristic increase of uric acid. Certain writers have attempted to connect a decreased elimination of uric acid with an accumulation of the same in the blood, giving rise to numerous disorders of which *gout* may be mentioned as one in which the connection has been, apparently, clearly shown. Great variations in the excreted uric acid seem to be characteristic of a train of disorders, rather than of a single one.

From recent investigations it appears that the ratio of excreted urea to uric acid is in health not far from 50:1, and that variations in this ratio are of greater moment than are variations in the absolute amount of the acid. Both must be considered as normal end products of nitrogenous metabolism, contrary to the older view that uric acid is the antecedent of urea, and that the amount of the former found in the urine represents merely that which failed to be completely oxidized. A marked change in the above ratio, 50:1, by increase of the uric acid is characteristic of a condition which is somewhat indefinitely called the uric acid diathesis.

In the recognition of uric acid the following points may be noted. When present in large amount it frequently precipitates from the urine in the free form, or as acid urates which have a yellowish color. When the amount present is small it may be found by acidifying with hydrochloric acid and then allowing the urine to stand some hours in a cool place; uric acid crystals separate. In mixed sediments it may be recognized by this test:

Murexid Test. Throw the sediment on a filter and wash once with water. Place the residue in a porcelain dish, add a drop of strong nitric acid, and evaporate to dryness on the water-bath. A yellow or brown mass is obtained, and this touched with a drop of ammonia water turns purple.

Unless the uric acid or urate is present in the sediment in fine granular form its recognition by the microscope is very simple. Illustrations of the forms of uric acid and certain urates are given in the chapter on the sediments.

The Amount of Uric Acid.

For the determination of the amount of the acid in the urine we have the choice of several methods, not one of which is very convenient or of great accuracy. The first of these depends on the fact referred to above, that hydrochloric acid liberates uric acid from its combinations, precipitating it in crystalline form.

Precipitation Test. Measure out 200 Cc. of urine and add to it 20 Cc. of strong hydrochloric acid. Mix thoroughly and set aside in a cool place for about 48 hours. At the end of this time collect the reddish-yellow deposit on a weighed filter, wash it with a little cold water dry and weigh. Not over 30 or 40 Cc. of water should be used in the washing. The precipitated uric acid is not pure, holding coloring and other substances which increase its weight. On the other hand, it is soluble to some extent even in cold acidulated water so that not the whole of it is obtained on the filter and a correction must be made. It is usually recommended to add to the weight obtained 4.8 Mg. for each 100 Cc. of filtrate and washings.

If the urine under examination contains albumin, the latter must be coagulated by heating with a drop or two of acetic acid and filtered out, before the test is made. If the urine is very cold to begin with and has a sediment of urates the latter must be brought into solution by warming before beginning the test. To prevent precipitation of phosphates during the warming a few drops of hydrochloric acid may be added. This method is at best only a rough approximation, but is the one by which most of our results have been obtained. The following gives better results:

Salkowski-Ludwig Method. The determination here is based on the fact that uric acid gives a very insoluble precipitate with ammoniacal solution of silver nitrate, from which precipitate after filtration and washing the acid may be readily separated, brought into *concentrated* solution, reprecipitated and weighed.

In using the method the following solutions are required.

(*a.*) **Ammoniacal Silver Nitrate.** Dissolve 25 Gm. of silver nitrate in 100 Cc. of distilled water, add ammonia water until the precipitate which appears at first is completely redissolved, leaving a clear solution. Make this up to 1,000 Cc. with distilled water and keep in a dark bottle or away from the light.

(*b.*) **Magnesia Mixture.** Made as described in the appendix. It must be strongly alkaline and clear, or nearly so.

(*c.*) **Solution of Potassium or Sodium Sulphide.** The pure crystals of sodium sulphide obtained from dealers in chemicals may be used by dissolving 25 to 30 Gm. (Na_2S, $9H_2O$) in 1,000 Cc. of distilled water. A solution may be made, also, by dissolving 10 Gm. of pure sodium hydroxide in 1,000 Cc. of water, and converting this into sulphide which is done as follows: Divide the solution into two equal portions. Saturate one thoroughly with hydrogen sulphide and to this then add the other half. Keep in a glass stoppered bottle, the stopper paraffined.

To make the test measure out 200 Cc. of the urine and transfer to a beaker. Add 20 Cc. of the silver solution, (*a*), to an equal volume of the magnesia mixture, (*b*), and then ammonia enough is added to clear up any precipitate which forms. This clear mixture is now poured into the urine in the beaker and the whole well stirred. A precipitate of silver urate forms along with silver and earthy phosphates. The excess of ammonia prevents the precipitation of silver chloride. Silver urate is quite insoluble in ammonia; it is gelatinous alone and does not settle very well but the phosphate precipitate corrects this difficulty to some extent. The beaker is allowed to stand at rest about an hour, after which the contents are filtered and the precipitate washed with weak ammonia on the filter. To do this the ammonia is sprayed into the beaker from a wash bottle and rinsed around thoroughly. This is done several times, the liquid being poured on the filter. Where available a Gooch crucible serves admirably for the collection of the precipitate as the filtration is slow on paper without aspiration. It is not necessary to remove any of the precipitate which clings to the beaker, as will be seen. When the washing is complete transfer the precipitate and filter paper, or asbestos if the Gooch crucible is used, back to the beaker and pour over it a boiling mixture of 20 Cc. of the sulphide solution, (c), and 20 Cc. of distilled water. Stir up thoroughly, allow to stand some time and then add 50 Cc. of boiling water. Place the beaker on a sand-bath or gauze and bring the contents to boiling, stirring continually. Keep hot some minutes and then allow to stand until cold, the precipitate being stirred meanwhile occasionally.

The treatment with the sulphide solution decomposes the silver urate with precipitation of black insoluble silver sulphide, the uric acid remaining in solution as soluble urate. The cooled liquid is filtered into a porcelain dish, and the precipitate washed with warm water, the washings going also into the dish. Enough hydrochloric acid is now added to combine with all the bases present and liberate the uric acid, which is the case when the liquid becomes acid in reaction. It is now slowly evaporated to a volume of about 10 Cc., best on a water-bath, and then allowed to stand an hour for the complete separation of the uric acid. This is then collected on a weighed Gooch crucible, the crystals being transferred gradually by aid of the filtered liquid. When the crystals are on the asbestos they are washed with a little acidulated water several times. The crucible is then dried at 100°, put back in the funnel and treated with a small amount of pure carbon bisulphide to remove traces of sulphur separated on decomposing the alkali sulphide. Finally, wash with ether, dry at 100° C. and weigh. The results are always a little low, but fairly regular.

As the acid is finally precipitated from a very small volume of liquid and but little water is used in washing no correction need be made for solubility, as in the first process described. While simple enough in principle and easily carried out considerable time is required for the performance of all the operations involved in the method.

The following method is free from this objection and is equally accurate:

Haycraft Method. This depends on the precipitation of the uric acid, as silver urate, and its subsequent titration by standard solution of ammonium sulphocyanate. The following solutions are required:

(*a.*) **Ammoniacal Solution of Silver Nitrate.** Dissolve 5 Gm. of crystals in 100 Cc. of water and then add enough ammonia water to give the solution a strong alkaline reaction. Make up to 200 Cc. with the ammonia.

(*b.*) **Ammonium Sulphocyanate.** This solution is made as described later for the determination of chlorides in urine by the Volhard method. It is given just one-fifth the strength there described and may be made by diluting 100 Cc. of that solution to 500 Cc. in a measuring flask.

(*c.*) **Ammonium Ferric Sulphate,** (ferric alum), saturated solution as indicator. Described under the chlorine test.

It has been shown by Dr. Haycraft that silver combines with uric acid in constant and definite proportion, viz., one atom of silver to one molecule of the acid, or 108 parts of silver to 168 of the acid, giving the formula $AgC_5H_3N_4O_3$.

This precipitate dissolves in dilute nitric acid and if the solution so obtained is treated with the ammonium sulphocyanate the following reaction takes place:

$$Ag\,C_5H_3N_4O_3 + NH_4SCN = Ag\,SCN + NH_4C_5H_3N_4O_3.$$

From this it follows that one cubic centimeter of a fiftieth normal, ($\frac{N}{50}$), solution of the sulphocyanate liberates and indicates .00336 Gm. of uric acid.

It is fully explained under the chlorine test that if a solution of a sulphocyanate is added to a solution of a silver salt containing nitric acid and ferric sulphate a complete reaction takes place between the sulphocyanate and silver before the characteristic reaction between the former salt and the ferric compound appears. In other words, the sulphocyanate and the silver combine first and then any further amount of sulphocyanate added unites with the iron, producing a red color (of ferric sulphocyanate) *indicating* the completion of the first reaction.

With these general explanations the process will now be understood.

Measure out 50 Cc. of the urine and warm it gently if it contains a sediment of urates. Add 3 to 4 Gm. of pure sodium bicarbonate and then ammonia enough to give a strong alkaline reaction. This may give a precipitate of phosphates which need not be heeded. Next add 5 Cc. of the silver solution, (*a*), and mix thoroughly. This produces a precipitate of silver urate along with the bulky phosphates thrown down by the ammonia. Allow to stand half an hour and then filter. A paper filter and funnel may be used in the usual manner, but much better results are obtained by the use of the Gooch crucible and asbestos with aid of an aspirator. Rinse the sides of the beaker thoroughly with weak ammonia and pour this on the precipitate in the funnel or crucible. Continue the washing of the precipitate with weak ammonia water until all traces of silver are washed out, as may be shown by allowing a few drops of the filtering washings to fall into some dilute hydrochloric acid in a test-tube. The washing is complete when a cloudiness is no longer obtained in this test.

Now pour some pure dilute nitric acid into the beaker in which the precipitation was made, and which was washed free from silver by the ammonia, and shake it around until any traces of the silver urate precipitate are dissolved. Put the funnel or Gooch crucible over a clean receptacle and pour this acid liquid on the precipitate. Silver urate dissolves completely in dilute nitric acid, and enough of this is added, a little at a time, to bring about complete solution. It now remains to titrate the silver in this solution. To this end add 5 Cc. of the ferric alum solution, and if the mixture is not clear and colorless, about 2 Cc. of pure strong nitric acid. Then from a burette run in the sulphocyanate, (b), a little at a time, shaking after each addition until a faint red shade of ferric sulphocyanate becomes permanent. Toward the end of the titration a red appears as each drop of liquid from the burette falls into the silver solution below, but this color fades out on shaking and does not persist until the last particle of silver has been taken up by the sulphocyanate. Supposing now that 15 Cc. of the latter solution are required to reach this point we have $15 \times .00336 = .0504$ Gm. as the amount of uric acid in the 50 Cc. of urine taken. A volume as large as this would seldom be required, 5 to 10 Cc., corresponding to 16.8 to 33.6 Mg., is usually sufficient.

The method gives results which are a little too high as the silver carries down traces of other bodies as well as uric acid. But the error is not great enough to interfere with the practical application of the process where even the best results are desired. The washing of the precipitate of silver urate is the point which requires the greatest care. A little practice will show how this can be best done.

Hippuric Acid.

This acid is found in very small amount normally in human urine, and is the chief nitrogenous product in the urine of the herbivora. It is increased in human urine by a diet of aromatic vegetable substances, but is never abundant enough to have clinical importance.

Chapter XV.

UREA.

UREA is the important nitrogenous substance excreted in human urine. A large part of the nitrogen of our food is normally converted into urea for elimination from the body. How this conversion takes place or where is not known. That a part, at least, *may* be formed in the liver from ammonium carbonate has been shown to be probable but the connection between it and the antecedent muscular tissue is still very obscure. Not far from 90 per cent of the nitrogen consumed as food is excreted as urea, but the absolute amount of the latter passed in a day is exceedingly variable. In the urine of the average man it is between 30 and 40 Gm. while in the urine of women it is less. The variations depend mainly on the diet, the urea being highest with a diet rich in meat, eggs, beans, peas and similar vegetables, and low with a diet of fruits, bread and potatoes. The *percentage* amount of urea depends further on the volume of the urine passed in a day and may vary from a change in the amount of water consumed and also from different losses by perspiration. The percentage amount of urea depends also on the time when the urine is voided. A determination of value should therefore be made on the mixed urine of the 24 hours.

It is usually assumed that 2 per cent is the average amount excreted in health, but this is probably low. While the variations from this mean are great in health they are much more marked in pathological conditions.

Urea is increased in total amount although it may be

diminished in percentage in diabetes mellitus and insipidus and also in fevers. It has been found to be increased in cases of poisoning by heavy metals, but why is not clearly demonstrated.

Clinically, the increase in diabetes and fevers is of the greatest interest because we have here evidence of increased consumption of the nitrogenous tissues of the body. A diminished elimination of urea has been observed in acute yellow atrophy of the liver and in other diseases of that organ. This has been taken to indicate that the liver may be the place of formation of urea. In cases of malnutrition in general the absolute and percentage amount of urea may be greatly diminished.

A marked decrease has been observed, also, in diseases involving structural changes in the tubules of the kidney as in parenchymatous nephritis.

Recognition of Urea. Because of its extreme solubility urea cannot be easily obtained by evaporation of urine. It has been shown, however, in an earlier chapter that by concentrating the urine slowly to a small volume— to one-third or one-fourth—cooling and adding strong nitric acid, a crystalline precipitate of plates of urea nitrate separates which is characteristic. From this precipitate pure urea can be obtained.

Clinically, this test has no importance as we are concerned only with a measurement of the amount of urea. This determination can be made in several ways, but in actual practice we employ three essentially different methods. The first depends on the fact that solutions of urea precipitate solutions of certain metals in a definite manner from which a volumetric process has been derived. The second depends on the fact that solutions of certain oxidizing agents decompose solutions of urea with the liberation of its nitrogen (and carbon dioxide) in gaseous form. From the known relations between weight and volume of the gas, and weight of nitrogen and weight of urea the absolute amount of the latter may be calculated. The third method depends on the fact that when the urea of urine is decomposed into water, carbon dioxide and nitrogen its specific gravity is decreased in a manner empirically determined. The loss in specific gravity bears a certain relation to weight of urea present.

DETERMINATION OF UREA.

Liebig's Method. We have here the oldest, and in many respects the best of our processes for the titration of urea. The principle involved in the method is this. When a solution of mercuric nitrate is added to a solution of urea a white precipitate forms and settles out. By working with solutions of a certain definite concentration it has been found that the reaction between the mercury and urea takes place in constant proportion and according to this equation:

$$2CON_2H_4 + 4Hg(NO_3)_2 + 3H_2O = $$
$$2CON_2H_4 \cdot Hg(NO_3)_2 \cdot 3HgO + 6HNO_3$$

This precipitate contains 10 parts of urea for every 72 parts of HgO. 72 Gm. of HgO dissolved in HNO_3 should precipitate, therefore, 10 Gm. of urea.

The same solution of mercury gives a yellow precipitate with solution of sodium carbonate which is used as an *indicator* in a manner to be described.

The urea solution to be analyzed is poured into a beaker and standard solution of the mercuric nitrate added gradually, with constant stirring from a burette. From time to time a drop of the liquid above the precipitate is taken on the end of a glass rod and brought in contact with a few drops of a concentrated solution of sodium carbonate on a plate of dark glass. A yellowish precipitate forms here if the drop contains any excess of the mercury compound beyond that necessary to precipitate the urea. The end of the reaction is frequently determined in this manner, as in the original process, but not with greatest accuracy. A modified process as now to be explained is preferable, and easily carried out.

As the equation above shows, nitric acid is set free in the reaction between the urea and the mercuric nitrate. This acid has a decomposing effect on the precipitate, tending to form new nitrate and thus diminish the amount which, theoretically, should be added for complete precipitation. The nitric acid must therefore be neutralized from time to time as formed, or better, just before the end reaction with the indicator is tried.

It has been found, also, that to precipitate exactly 10 Mg. of urea in this manner, not 72 Mg. of mercuric oxide in solution, but a slightly greater amount must be used. The experiments of *Pflueger* showed that 77.2 Mg. is needed for the purpose and the standard solution should be made to contain 77.2 Gm. per liter.

In the titration of urine certain modifications must be made which are not necessary in the titration of pure urea solutions. The phosphates, sulphates and chlorides of urine interfere with the reaction and must be removed before the test is begun.

The phosphates and sulphates may be removed by precipitation with barium solution, while the chlorides may be thrown out by silver nitrate. It is also possible to make a correction for the chlorides instead of precipitating them. The following solutions are necessary in making the test.

(*a*). **Mercuric Nitrate Solution.** This is made of definite strength and should contain the equivalent of 77.2 Gm. of the oxide in one liter. In making this solution we may start with pure metallic mercury, with mercuric oxide or with the commercial nitrate (mercurous). With mercury it can be made in this manner: Weigh out a quantity of pure mercury and heat it in a porcelain dish or casserole with two to three times its weight of strong nitric acid of 1.42 Sp. Gr. When the mercury is in solution evaporate to the consistence of a thick syrup and add from time to time a few drops of nitric acid to complete the oxidation. When the addition of the acid is no longer followed by the evolution of red fumes the action is complete and the mercury exists as mercuric salt. Now pour into the syrupy residue ten times its volume of water with constant stirring. In adding the water it always happens that a little of the nitrate is decomposed and thrown down as a basic salt. Allow the liquid to settle thoroughly, pour off from the sediment and dissolve the latter in a few drops of nitric acid. Add this now to the main solution and dilute it with distilled water to make 1 liter of each 71.5 Gm. of mercury.

When mercuric oxide is employed, weigh out the proper amount, dissolve it in a slight excess of strong, pure nitric acid, evaporate to a syrup and treat as above. Finally, dilute with water to yield a solution with 77.2 Gm. to the liter.

(*b*). **Baryta Solution** to precipitate phosphates and sulphates. To one volume of a cold saturated solution of barium nitrate add two volumes of a cold saturated solution of barium hydroxide. Keep in a well stoppered bottle.

(c). **Sodium Carbonate Solution.** This is best made of the pure, dry carbonate readily obtained as a commercial article. It must be remembered, however, that the so-called dry carbonate contains a little water, which may be removed by heating, in a platinum dish, to low redness. Dissolve 53 Gm. of the salt thus dried, in water and dilute to one liter.

A mercury solution made of pure material according to the above directions should have the correct strength, but for control it may be tested by means of a solution of pure urea in water.

(d). **Standard Urea Solution.** Weigh out 2 Gm. of pure urea, which is now readily obtained, and dissolve it in distilled water to make 100 Cc.

Before making the test proper it is necessary to determine the relation of the sodium carbonate solution to the mercury solution in presence of urea. This may be done by taking exactly 10 Cc. of the urea solution and adding to it 19 Cc. of the solution of mercuric nitrate. Shake thoroughly, allow to stand a minute and filter. Wash the precipitate with a little distilled water, and to the mixed filtrate and washings add two drops of a weak solution of methyl orange, or enough to give a pink color. Then from a burette run in the solution of sodium carbonate, with constant shaking, until the pink color changes to yellow. The reaction is sharp enough for the purpose. Not over 11.5 Cc. of the alkali solution should be required for this. Calculate the amount needed for each Cc. of mercuric nitrate. Proceed now with the actual test.

Measure 10 Cc. of the standard urea solution again and run in 19.5 Cc. of the mercuric nitrate. Add now the correct number of cubic centimeters of soda solution required to neutralize the acid from the nitrate as calculated from the results of the last experiment. Then by means of a stirring rod bring a drop of the liquid in the beaker in contact with a drop of a semifluid mixture of sodium bicarbonate and water on a dark glass plate. Stir the two together and observe the color. It should be white. Run in two or three drops more of the mercury solution, stir well and repeat the test, and continue until a slight yellow color is obtained on mixing the drops from the beaker with the moist bicarbonate. If the mercury solution is correct just 20 Cc. should be used for this.

The sodium bicarbonate used as indicator must be pure, especially as regards freedom from chloride. An excess of it can be washed in a beaker several times with a little cold water, which is poured off leaving the salt finally in a pasty condition suitable for use.

We proceed now to the actual test of a sample of urine. Measure off accurately a definite volume and add to it just half its volume of the baryta solution. Convenient proportions are 50 Cc. of urine and 25 Cc. of the baryta solution. Shake thoroughly and filter through a dry filter into a flask. Measure out now exactly 15 Cc. of this filtrate which contains the urea of 10 Cc. of the original urine, the baryta solution having taken out only phosphates, sulphates and carbonates, with certain bases. This filtrate still contains chlorides which are objectionable and which could be separated by another precipitation with the proper amount of silver nitrate. It will be, however, more convenient and fully as accurate to make a correction for them at the end of the test, as will be explained. The 15 Cc. of filtrate has an alkaline reaction and must be neutralized. (If not alkaline a new precipitation must be made, taking equal volumes of urine and baryta solution, and finally 20 Cc. of the filtrate.) The neutralization can be effected by adding carefully, a drop at a time, dilute nitric acid, testing with litmus paper.

The filtrate thus prepared is titrated with the mercury solution. Begin by adding a Cc. at a time and after each addition bring a drop of the mixture in contact with a drop of the semifluid sodium bicarbonate on a plate of dark glass. The drops should be placed side by side and mixed at the edges. At first the mixture remains white, even after stirring, but as the addition of mercury is continued a point is reached where the drop from the beaker brought in contact with the moist bicarbonate gives a light yellow shade. On stirring the drops together this yellow should disappear, but this shows that the end of the reaction is nearly reached. Add now the mercury solution in drops and test after each addition. When the point is reached where a faint yellow shade persists after stirring together the drop from the beaker and the sodium bicarbonate it is time to neutralize with the normal sodium carbonate solution. Run in the right number of cubic centimeters corresponding to the mercury used and now make the test for the final reaction again and continue until the yellow color appears.

Regard this test as preliminary and make a new one with 15 Cc. of the filtrate neutralized as before. Run in directly within 1 Cc. of the amount of mercury required, as shown by the first test, neutralize and complete as before. For each cubic centimeter used, *after deducting for chlorides*, calculate 10 Mg. of urea. The correction for the chlorides is based on the following principle : In the presence of sodium chloride, or other chloride, the reaction between mercuric nitrate and urea does not begin until enough of the former has been added to form mercuric chloride with all chlorine present, according to the following equation:

$$Hg(NO_3)_2 + 2NaCl = 2NaNO_3 + HgCl_2$$

For the nitrate corresponding to 216 parts of HgO we use here 117 parts of salt, or for 117 milligrams of salt, 216 milligrams of mercuric oxide, or 2.79 Cc. of the standard mercuric nitrate solution.

One milligram of sodium chloride, therefore, combines with the mercury compound in .0238 Cc. of the standard solution. Mercuric chloride does not react with the sodium bicarbonate, and the amount formed is beyond that shown by the titration. Therefore, to apply the correction, determine the chlorides present in 10 Cc. of urine (by a process to be given later) calculate to sodium chloride, and for each milligram found deduct .0238 Cc. from the volume of the mercuric nitrate solution used in the titration. As the amount of chlorine in the urine is about equivalent to one per cent of salt in the mean, an approximate correction is often made by subtracting 2 Cc. from the volume of the mercury solution.

The above calculations are based on the supposition that the original urine contains 2 per cent of urea, and that a volume of about 20 Cc. of mercuric nitrate is used in the titration. If the per cent of urea is much greater or less than this a correction on account of volume must be made. This correction has been worked out empirically by Pflueger, and, without discussing how it is derived, it will be sufficient to explain its application.

If more than two per cent of urea is present it is necessary to add more than 20 Cc. of the mercuric solution in titration. If the volume of the latter solution is greater than the sum of the volumes of the prepared urine and soda solution used in neutralization, this sum must be subtracted from the volume of the mercury solution and the result multiplied by 0.08. The product is added to the number of cubic centimeters of mercuric nitrate used, to give the corrected result. If, on the other hand, the volume of mercuric nitrate used in titration is less than the sum of the volumes of prepared urine and soda solution the difference is multiplied by 0.08 and the product taken from the number of Cc. of mercuric solution used, to give the corrected result. In these calculations the volume of mercuric nitrate taken up by the chlorides must be considered as part of the diluting liquid. The same must be remembered in adding sodium carbonate for neutralization.

The correction may be expressed in this formula, according to Pflueger:

$$C = -(V_1 - V_2) \times 0.08,$$

in which

C = the correction to be added or subtracted.

V_1 = the sum of the volumes of the urine, soda solution and mercuric nitrate combined with the chlorides.

V_2 = the volume of mercuric nitrate taken by urea.

In illustration we may take an actual case:

15.0 Cc. = the prepared urine (neutralized).
14.8 Cc. = the sodium carbonate used.
1.8 Cc. = the mercuric solution used by chlorides.

V_1 = 32.6 Cc.
V_2 = 26.0 Cc.
  6.6

$$-(V_1 - V_2) \times 0.08 = -0.528 = C.$$

Therefore, 26—0.5 = 25.5 is the corrected volume of mercuric nitrate, indicating, with the latter solution of standard strength, 25.5 Gm. of urea in a liter.

Method by Liberation of Nitrogen.

—A solution of urea is decomposed by a solution of a hypochlorite or hypobromite as illustrated by this equation.

$$CON_2H_4 + 3NaOCl = CO_2 + N_2 + 2H_2O + 3NaCl.$$

That is, nitrogen and carbon dioxide gases are given off. If the reaction is allowed to take place in an alkaline medium the carbon dioxide will be held and the nitrogen alone given off. The volume liberated is a measure of the weight of urea decomposed. From the above equation it is seen that 28 parts by weight of nitrogen correspond to 60 of urea, from which follows that one cubic centimeter of pure nitrogen gas, measured at a temperature of 0°C and under the normal pressure of 760 Mm. corresponds to .00269 Gm. of urea. One Gm. of urea furnishes 371.4 Cc. of nitrogen gas.

In employing these principles in practice it is simply necessary to bring together a measured volume of the urine or urea solution and the hypochlorite or hypobromite reagent under such conditions that all of the nitrogen liberated may be collected and accurately measured.

As a reagent a solution of sodium hypobromite is very commonly employed. As it does not keep well it must be made fresh for use, which is inconvenient unless many tests have to be made at one time. The reagent may be prepared in this manner:

Dissolve 100 Gm. of good sodium hydroxide in 250 Cc. of water. When cold add 25 Cc. of bromine by means of a funnel tube carried to the center of the solution. The bromine must be poured into the funnel, a little at a time, and the lower end moved around to act as a stirrer and mix the liquids. During the reaction the bottle or flask containing the alkali should be surrounded by cold water, and the mixture should be made out of doors or in a good fume closet. The finished solution contains an excess of alkali sufficient to hold the carbon dioxide given off when it reacts on urea. In place of this solution a strong hypochlorite solution may be used with advantage. The U. S. P. solution *when properly made* answers very well. Its preparation is given in the appendix.

Many forms of apparatus have been devised for this purpose, one of the oldest and best known of which is that of Huefner, shown in Fig. 42.

At A below is a small vessel, the bottom of which rests on the support, and which holds the urine to be decomposed. The capacity of this vessel is usually between 5 and 10 Cc., but must be accurately determined by filling with mercury, pouring this out and weighing it. Above A, and separated from it by a ground glass stopcock, is the much larger vessel B, which holds the hypochlorite or hypobromite reagent. On the narrow neck of B rests a cup-shaped receptacle, C, to hold water. Finally, over B a measuring tube is supported in

FIG. 42.

such a manner that it must receive any gas passing up from *B*. At the beginning of the experiment this measuring tube *D* is filled with water.

The apparatus is used in the following manner: Rinse out *A* and *B* and pour into the latter more than enough urine to fill *A*. Open the stopcock and allow the urine to flow down, which may be assisted when slow by moving a thin glass rod or bit of wire up and down through the opening in the stopper. When *A* is quite full close the stopper, rinse the urine from *B* and fill it with the hypobromite reagent. Now fill the cup *C* with water so that its surface is one or two centimeters above the opening into *B*, fill the graduated tube with water, close the end with the thumb and invert it in the cup. Finally fasten it in position over *B* as shown.

The stopcock is now opened which permits the heavier reagent to sink and slowly mix with the urine. A liberation of gas soon begins and proceeds slowly. At the end of twenty or thirty minutes the reaction is complete, which can be noted by the disappearance of the gas bubbles. Close the end of the gas tube with the thumb, remove it and immerse in a jar of water, having the temperature of the air, until the water levels inside the tube and in the jar, are the same. Clamp the tube in position and allow it to stand until the temperature of the gas becomes constant, which may require fifteen minutes. Finally adjust the tube level again if necessary, read the volume of the gas, note the height of the barometer and the temperature as given by a thermometer hanging near the tube and reduce this volume to standard conditions by the following formula:

$$V_c = V_o \cdot \frac{b-w}{(1+.00366t)760}$$

In this formula

V_c = the corrected volume.
V_o = the observed volume.
b = the barometric height.
w = the tension of water vapor at the observed temperature, in millimeters of mercury.
t = the temperature as observed.

The values for w at different temperatures are given in this table.

t	w	t	w
10°	9.165 Mm.	18°	15.357 Mm.
11°	9.792 Mm.	19°	16.346 Mm.
12°	10.457 Mm.	20°	17.391 Mm.
13°	11.162 Mm.	21°	18.495 Mm.
14°	11.908 Mm.	22°	19.659 Mm.
15°	12.699 Mm.	23°	20.888 Mm.
16°	13.536 Mm.	24°	22.184 Mm.
17°	14.421 Mm.	25°	23.550 Mm.

Having thus the volume of the gas under normal conditions, the weight of urea corresponding can be calculated by data already given.

The results obtained by this method are too low and must be corrected, as will be explained later.

Another form of urea apparatus which can be constructed by any one is shown in Fig. 43.

The tall jar is filled with water which must stand until it has the air temperature. A 50 Cc. burette is inverted in the jar, the delivery end being connected with a bottle holding about 150 Cc. by means of a piece of firm rubber tubing. The rubber tube is slipped over a short glass tube passing through the hole in a rubber stopper which must close the bottle accurately. In the bottle is a short stout test-tube, or vial, holding about 10 Cc. and which contains the urine to be tested. Into the bottle itself is poured the reagent as above described, which must not reach to the top of the test-tube. On mixing the liquids the urea decomposes, liberating the gas as before, which passes through the rubber tube and displaces water in the burette so that the volume can be readily determined.

The test is made practically in this manner: Pour about 20 Cc. of the strong hypobromite or 40 Cc. of the weaker hypochlorite reagent into the bottle. With a pipette measure some exact volume of urine, usually 5 Cc., into the small test-tube, and by means of small iron forceps place the latter carefully in the bottle containing the reagent. Next insert the stopper which connects the bottle with the burette standing in the jar of water. The level of the water in the burette is displaced by this operation. Now allow the apparatus to remain at rest ten minutes until the air in the mixing bottle and tube has the temperature of the outside air. Through handling it of course becomes warmer. At the end of the time, by means of the attached clamp and not by the hand, lift the burette, or depress it if necessary, until the water levels inside and outside are the same. Note the position of the water on the burette graduation. Read the air temperature by a thermometer, which should be suspended near the tube, and observe the height of the barometer.

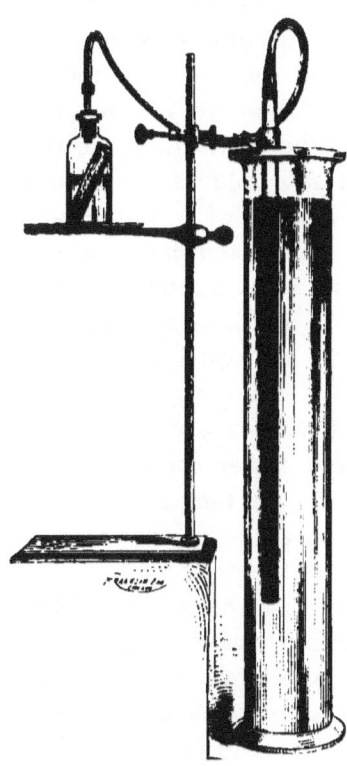

FIG. 43.

Incline the bottle to mix the urine and the reagent,

shake gently and repeat these operations from time to time. On the decomposition of the urea nitrogen gas or its equivalent volume of air passes over into the burette and forces out some of the water. When, after repeatedly shaking the bottle, no increase of the gas volume in the burette can be observed allow the whole apparatus to stand until the contents of the bottle and burette have cooled down to the air temperature again. Then lift the burette, with the clamp as before, to restore the levels and read the gas volume. From this substract the volume at the first reading. The difference is the volume of nitrogen gas liberated in the reaction, at the observed temperature and atmospheric pressure. If, as sometimes happens, more gas is liberated than the burette will hold, repeat the experiment, using urine diluted with an equal volume of water.

The calculations are made as in the case given above, as the gas volume is finally measured under the same conditions. It is assumed here that the air temperature and barometric pressure remain constant during the experiment.

Exact investigations have shown that the whole of the nitrogen is not liberated in the reaction, as at first assumed, but falls short between seven and eight per cent. It appears that four or five per cent of the nitrogen of the urea is oxidized to nitric acid in the reaction and escapes measurement. Another small portion is left in the ammoniacal condition. Attempts have been made to prevent this abnormal oxidation by adding to the urine a reducing agent to destroy nitric acid as fast as formed. Dextrose has been used for the purpose, also cane sugar, and apparently with success. But a great excess of sugar must be added.

It has been shown, also, that the theoretical yield of

nitrogen can be approached by a modified method of applying the reagent. *J. R. Duggan*, working in *Remsen's* laboratory, found that by mixing the alkali with the urine and then adding the bromine so as to form hypobromite in presence of the urea the yield of nitrogen is much increased, reaching nearly the amount called for by theory. The simple bottle and burette apparatus may be used in this manner by measuring the alkali, 20 Cc. of 20 per cent sodium hydroxide solution, and urine, 5 Cc. into the bottle, and the bromine, about 1 Cc., into the test-tube. The mixture is made and process completed as before. The modification has not come into general use, probably because of the objection to working with free bromine, at each test. As the deficiency by the usual method has been shown to be nearly eight per cent, it is sufficient for most purposes to assume that the volume of gas obtained is 92 per cent of the whole and correct by calculation.

It remains now to describe two forms of apparatus which are used without any correction, for rapid clinical tests.

The Squibb Apparatus is the first of these, and one which can be very highly recommended. The construction of the apparatus is shown by Fig. 44.

The upright bottle to the left contains the reagent, hypochlorite or hypobromite as before. By means of a pair of small forceps a short test-tube, F, containing a measured volume of the urine is dropped into the bottle, but carefully so that the liquids will not mix until the bottle is shaken. A bent glass tube and a rubber tube connect this with the bottle B, which at the beginning of the test is quite filled with water. Another glass tube connects B with the bottle D, empty at the beginning of the test.

To make the test pour into the first bottle 20 Cc. of strong hypobromite solution, or 40 Cc. of the hypochlorite. Measure accurately 4 or 5 Cc. of urine into the tube, drop this into the bottle and insert the stopper. Fill B quite full of water, and insert its stopper which drives out a little water through the short tube into D. Allow the whole apparatus to stand ten minutes to take the temperature of the air, then empty D and replace it. Now tip the first bottle so as to mix the contents of F with the reagent

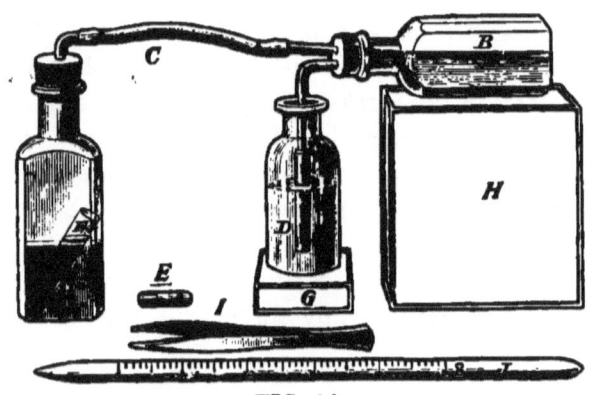

FIG. 44.

and shake gently. Bubbles of gas escape and passing over into B drive out a corresponding volume of water. Repeat the shaking of the reagent bottle several times. In a few minutes the reaction is complete, but the apparatus must be allowed to stand to cool down to the air temperature. A part of the water in D may be drawn back into B. Finally measure the volume of water left in D, and take this as the volume of gas liberated. Make the calculation as before on the assumption that each Cc. of gas corresponds to .0027 Gm. of urea. In this there are two errors which nearly compensate each other. In the first place not all the gas is liberated for the reasons explained above, but in the second place the volume read off is higher than normal because of higher temperature and lower barometer. The results obtained are therefore nearly correct, sufficiently so for all clinical purposes.

Squibb has constructed a table from which the percentage of urea corresponding to any volume of gas, for a given volume of urine taken, can be read at a glance without any calculation whatever. This is convenient, but not necessary.

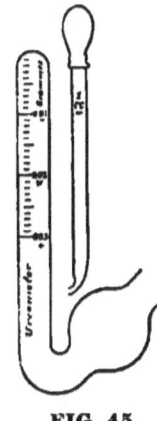

FIG. 45.

The Doremus Apparatus. This is shown in the annexed cut, and at the present time is very widely used by physicians because of the simplicity of the method of employing it.

The graduated tube is filled to an indicating mark with strong hypobromite solution and then water is added to fill the remainder of the tube and the bulb. By means of the pipette, graduated to hold 1 Cc., this quantity of urine is forced into the liquid in the upright part of the tube. The urea is immediately decomposed, and its nitrogen ascends to the top of the graduated part, where it is read off. The longer marks on the tube indicate the fraction of a gram of urea in the one cubic centimeter of urine taken; the shorter marks indicate tenths. Multiplying by 100 we obtain the number of grams of urea in 100 Cc. The results are apt to be low from a loss of nitrogen through the bulb which can scarcely be avoided. The instrument is said to be "good enough" for clinical purposes, but cannot be compared with that of Squibb for accuracy. No instrument in which the volume of urine taken for the test is as small as one cubic centimeter should be expected to give even approximately accurate results, unless very great precautions are taken in the measurement and in subsequent parts of the work.

Chapter XVI.

THE DETERMINATION OF PHOSPHATES AND CHLORIDES.

Phosphates.

PHOSPHORIC ACID occurs normally in the urine combined with alkali and alkali-earth metals of which combinations the alkali phosphates are soluble in water, while the earthy phosphates are insoluble. In the urine, however, they are held in solution through several agencies. The larger part of the earthy phosphates appear to be held here normally in the acid condition, that is as compounds of the formulas $CaH_4(PO_4)_2$ and $MgH_4(PO_4)_2$. The salts of the type $CaHPO_4$ are present, also, in small amount. As long as the urine maintains its acid reaction these bodies may be expected to remain in solution, but if it becomes alkaline by fermentation, or by the addition of the hydroxides or carbonates of ammonium, sodium or potassium, the acid phosphates are converted into insoluble, neutral phosphates and precipitated. Most urines contain along with the acid phosphate traces of neutral phosphates which precipitate on boiling. It has been suggested that these phosphates are held by traces of ammonium compounds or by carbonic acid, both of which are driven off by heat, allowing the phosphates to precipitate. It is well known, however, that some urines can be boiled without showing any sign of precipitation. In such cases it is probable that the neutral phosphates are not present.

Part of the phosphoric acid of the urine comes directly

from the phosphates of the food and another portion results from the oxidation of the phosphorus holding tissues. In health the rate of such oxidation is practically constant, or nearly so, but in disease it may be greatly increased or diminished. Variations in the amount of excreted phosphates may become, therefore, of considerable clinical importance. Great care must be observed, however, in drawing conclusions regarding the destruction of phosphatic tissues from the results obtained by analysis, because of the uncertainty of the amount taken with the food and of what must be considered a normal excretion of phosphoric acid.

Wheat bread constitutes one of our important articles of food containing a relatively high amount of phosphates. Owing to changes in milling processes introduced and extended in the past twenty years very material reductions have been made in the percentage of phosphates and other mineral substances left in our fine flour. The resulting diminution in the phosphates of the urine from this cause is appreciable. It must be remembered also that a part of the phosphoric acid taken with the food is eliminated in insoluble form with the fæces. The amount so disposed of depends on the nature of the original condition of combination of the phosphoric acid and on the amount of alkali-earth bases present. These tend to form insoluble phosphates. With an exclusively vegetable diet the phosphates would, for this reason, be low, while with a meat diet more would be excreted, because here the alkali phosphates, especially potassium phosphate, are in excess. On the other hand the phosphates are greatly increased in the urine of people who consume great quantities of the various phosphatic beverages which have become popular in the United States in the past few years. It is readily seen how quite erroneous conclusions could be drawn from the tests of urine of such persons. Single analyses for phosporic acid may be very misleading. When the amount of phosphoric acid passed with a given diet is known, variations observed may sometimes be traced to certain pathological conditions briefly mentioned in the following paragraph.

In disease phosphates are found increased in *rickets* and *osteomalacia*, possibly from the failure to deposit the earthy phosphates in the bones. Meningitis is accompanied by an increase of the phosphates. The same is true of other disorders of the brain as this organ is especially rich in phosphatic substances. It is said that phosphates are increased after a period of severe nervous strain. What has been termed *phosphatic diabetes* has been described as a condition in which a persistent excretion of relatively large amounts of the phosphates is observed. The earthy phosphates so discharged may amount to 20 grams or more in a day. The causes leading to this condition are not clearly defined.

The phosphates have been found diminished in diseases accompanied by diminished nutrition, and in some diseases of the heart and kidney. In dilute urine, low specific gravity and high volume, the *percentage* amount of phosphoric acid is much decreased, but it does not follow that a decrease in the amount excreted in the twenty-four hours must also be small. A safe estimate can be made only with a mixed specimen taken from the urine of the whole day with consideration of the volume passed.

Various statements are found in the books regarding the mean excretion of the alkali and earthy phosphates. Different observers have reported between 2 and 5 Gm. of phosphoric anhydride (P_2O_5), while 3 Gm. may be taken, perhaps, as the mean.

The recognition of the phosphates is an extremely easy

matter. The presence of earthy phosphates may be shown by adding to the urine enough ammonia water to give a faint alkaline reaction and then warming. A flocculent precipitate, resembling albumin, appears and is usually white, or nearly so. But sometimes coloring matters come down with it in amount sufficient to give it a brownish or reddish shade. It will be recalled that the color of this precipitate was referred to under the head of blood tests.

The alkali phosphates can be detected in the filtrate after separation of the earthy phosphates. To this end add to the clear alkaline liquid a little more ammonia and some clear magnesia mixture. A fine crystalline precipitate of ammonium-magnesium phosphate separates and settles rapidly. This is very characteristic. The qualitative tests for phosphates have, however, little value in examination of the urine. We are chiefly concerned with the amount, the measurement of which will now be described.

Determination of Phosphates. It is customary to measure the total phosphoric acid, not the alkali or earthy phosphates separately. We have at our disposal several methods, gravimetric and volumetric, of which the latter are accurate and most convenient. A volumetric process will be described which serves for the measurement of the phosphoric acid as a whole, and which can be used for the separate measurement of the earthy and alkali phosphates by dealing with the precipitate and filtrate described in the qualitative test above. This method depends on the fact that solutions of uranium nitrate or acetate precipitate phosphates in greenish yellow colored, flocculent form, and that in a solution holding in suspension a precipitate of uranium phosphate any excess of soluble uranium compound may be recognized by the reddish brown precipitate which it gives with a solution of potassium ferrocyanide. The latter substance serves, therefore, as an indicator. If to a phosphate solution in a beaker a dilute uranium solution be added precipitation continues until the whole of the phosphates have gone into combination with the uranium. If, during the precipitation, drops of liquid from the beaker are brought in contact with drops of fresh ferrocyanide solution on a glass plate no reddish brown precipitate of uranium ferrocyanide appears until the last trace of uranium phosphate has been formed. The production of uranium ferrocyanide is the indication, therefore, of the finished precipitation of the phosphates.

The reaction between uranium and phosphates in acetic acid solution is illustrated by this equation:

$$UO_2(NO_3)_2 + KH_2PO_4 = UO_2HPO_4 + KNO_3 + HNO_3$$

From this it appears that 239 parts of uranium are required to precipitate 71 parts of P_2O_5. In order to have the reaction take place as represented above, it is necessary to neutralize the liberated nitric acid as fast as formed or dispose of it in some other manner. The best plan is to add to the solution to be precipitated acetic acid and *sodium acetate*, the first of which brings the phosphates into the acid condition, while the second is decomposed with the formation of sodium nitrate and free acetic acid. Mineral acids interfere with the reaction, while moderate amounts of acetic acid do not.

In order to carry out this method we prepare the following solutions:

(*a*). **Standard Uranium Solution.** This is made by dissolving 36 Gm. of the pure crystallized nitrate, $UO_2(NO_3)_2 6H_2O$ in distilled water to make one liter. The strength of the solution is adjusted by experiment as explained below.

(b). **Standard Phosphate Solution.** This is made by dissolving 10.085 Gm. of pure crystals of sodium phosphate, $HNa_2PO_4.12H_2O$, in distilled water to make one liter. 50 Cc. of this solution contains .100 Gm. of P_2O_5. Small fresh, uneffloresced crystals of the phosphate must be used for this solution.

(c). **Sodium Acetate Solution.** Dissolve 100 Gm. in 800 Cc. of distilled water, add 100 Cc. of 30 per cent acetic acid and then water enough to make one liter.

(d). **Fresh Ferrocyanide Solution.** Dissolve 10 Gm. of pure potassium ferrocyanide in 100 Cc. of distilled water. The solution should be kept in the dark.

The actual value of the uranium solution is determined by the following experiment: Measure out 50 Cc. of the phosphate solution, (b), add 5 Cc. of the acetate solution, (c), and heat in a beaker in a water-bath to near the boiling temperature. Place several drops of the ferrocyanide solution on a white plate. Fill a burette with the uranium solution and when the solution in the beaker has reached the proper temperature run into it from the burette 18 Cc. of the uranium standard. Warm again, and by means of a glass rod bring a drop of the liquid in the beaker in contact with one of the ferrocyanide drops on the plate. If the uranium solution has been properly made no red color should yet appear. Now run in a fifth of one Cc. more from the burette, warm and test again, and repeat these operations until the first faint reddish shade begins to show on bringing the two drops in contact. With this test as a preliminary one make a second, adding at first one-fifth of a cubic centimeter less than the final result of the preliminary, and finish as before. Something less than 20 Cc. should be needed to complete the reaction. Supposing 19.8 Cc. are required for the purpose the whole solution should be diluted in the proportion,

$$19.8 : 20 :: a : x$$

in which a represents the volume on hand. We have now a standard uranium solution, each cubic centimeter of which precipitates exactly 5 milligrams of P_2O_5, and with this we are able to measure phosphoric acid in *unknown* solutions.

It is sometimes recommended to use uranium acetate instead of nitrate in making this standard solution, and so avoid a disturbing element in the liberation of the nitric acid. But there are several advantages in the use of the nitrate which should be mentioned. In the first place its solutions keep better, and secondly, it can be obtained in commerce in almost (if not quite) chemically pure condition, so that it is possible to make up a solution of nearly correct strength by simply weighing out and dissolving the crystals. Assuming 239 as the atomic weight of uranium, and O$=$16, the relation of the crystallized pure nitrate to P_2O_5 is

$$1006 : 142,$$

from which it follows that a liter of the standard solution should contain 35.4 Gm. of the uranium salt, if each cubic centimeter is to indicate five milligrams of P_2O_5.

The solution of sodium acetate with acetic acid must always be added in the proportion given above if uniform results are expected, and the ferrocyanide indicator must be fresh and as weak as given.

The test of urine is made exactly as given in the above. Measure out 50 Cc. of urine, add 5 Cc. of the acetate mixture, and finish as before. The 50 Cc. of urine, in the mean, contains about as much phosphoric acid as was present in the same volume of standard phosphate solution. The titration must be made hot, because the reaction is much quicker and sharper in hot solution than in cold. Make always two tests; the first is an approximation, while the second gives a much closer result.

A separate test of the earthy phosphates may be made by adding to 200 Cc. of urine enough ammonia to give an alkaline reaction. The urine then must stand until the precipitated phosphates settle out. The precipitate is collected on a small filter, washed with water containing a very little ammonia, and then allowed to drain. It is next dissolved in a small amount of acetic acid, the solution diluted to 50 Cc., mixed with 5 Cc. of the sodium acetate solution and titrated as before. The reaction here is not quite as accurate as with the alkali phosphate, but the results are satisfactory for the purpose.

The difference between the total phosphates and the earthy phosphates, expressed in terms of P_2O_5, is the amount combined as alkali phosphates.

The Determination of the Chlorides.

Practically, all of the chlorine taken into the stomach with the food is eliminated with the urine. The chlorine entering the body is mostly in the form of sodium chloride, although traces of other chlorides are found in some of our foodstuffs.

The amount of salt excreted depends, therefore, closely on that consumed, and varies within wide limits. It is commonly said that 10 to 15 Gm. daily include the amounts passed in the great majority of urines. It must be remembered, however, that in individual cases the upper limit may be very greatly exceeded. In the urine of men who eat a great deal of salt food from 20 to 25 Gm. is frequently found. In such cases the volume of water drunk is usually large, so that the *percentage* amount of salt passed does not follow the same variations. Expressed in this manner, about one per cent represents the average excretion.

Pathologically, chlorides are increased in total amount in *diabetes insipidus*, and temporarily, sometimes in intermittent fevers. A decrease in the chlorides is more frequently observed, and has greater clinical importance.

This decrease is noticed generally in febrile conditions, and especially if salty exudations are being formed in any part of the body. In the serous accumulations of pleurisy common salt is abundant, and at the same time greatly diminished in the urine. The expectorated fluid in cases of acute pneumonia contains much salt, which in consequence is decreased in the urine. In some cases the chlorine is practically absent from the urine. In any event its reappearance in normal amount during the progress of disease is a favorable sign, as indicating the approach of normal conditions of absorption and excretion in the body. Quantitative tests for the chlorides of the urine become, therefore, of great importance, as their absence or marked decrease can only occur in disorders of serious nature. Fortunately, such tests are very easily made, and by simple volumetric processes.

Volumetric Determination of Chlorine. The best processes by which chlorine is measured volumetrically in the urine and elsewhere depend on the reaction between chlorides and silver nitrate,

$$NaCl + AgNO_3 = AgCl + NaNO_3,$$

from which it appears that 5.84 milligrams of salt require

for precipitation 16.95 milligrams of silver nitrate. In many cases if a weak solution of silver nitrate be added to a weak solution of salt in a bottle, with frequent shaking, the curdy precipitate of silver chloride formed will settle out so rapidly that it is possible to determine just when the reaction is complete from the formation of no further precipitate in the nearly clear liquid as drops of silver nitrate mix with it. A chloride can be added to precipitate silver from a solution of its nitrate in the same manner, and so delicate is the reaction that a method based on it is still employed in many mints for determining the amount of silver in bullion, coin or other alloy.

For the determination of chlorides in solutions, especially in urine, other methods, more convenient and fully as accurate, are employed. If a weak solution of silver nitrate be added to a neutral solution of a chloride containing enough potassium chromate to impart a slight yellow color to it the silver combines with the chlorine first and then, only after this has been completely precipitated, with the chromic acid to form brick-red silver chromate. In such a mixture the appearance of the first tinge of red is the indication that the chlorine has been wholly precipitated. The neutral chromate is used here as the indicator. As each drop of silver nitrate solution falls into the solution of chloride and chromate a transitory reddish color may appear before the reaction is completed, but this vanishes on shaking or stirring the liquid and becomes permanent only when the chlorides are completely combined with silver. In using this method the following solutions are employed:

(a.) **Standard Silver Nitrate Solution**, $\frac{N}{10}$. Dissolve 16.954 Gm. of pure fused silver nitrate in distilled water and dilute to one liter. Silver nitrate can usually be obtained of sufficient purity for the purpose, from dealers in fine chemicals, but should be fused in a porcelain crucible at a low temperature before being weighed out. The correctness of the solution may be tested by the following:

(*b.*) **Standard Sodium Chloride Solution,** $\frac{N}{10}$. Dissolve 5.837 Gm. of pure, dry, recrystallized sodium chloride in distilled water and dilute to one liter.

(*c.*) **Potassium Chromate Indicator.** Dissolve 10 Gm. of pure crystals, free from traces of chlorine, in 100 Cc. of distilled water.

To test the accuracy of the standard silver solution fill a burette with the same and then measure into a beaker 25 Cc. of the salt solution and add a few drops of the indicator. Now slowly run solution from the burette into the beaker, shaking the latter meanwhile, and continue until the curdy white precipitate shows a tinge of red from presence of a little chromate formed. Exactly 25 Cc. of the silver solution should be needed for this. One Cc. of the silver solution so made precipitates 3.537 Mg. of chlorine or shows 5.837 Mg. of sodium chloride.

This method cannot be applied directly to the urine because of its yellow color which obscures the end reaction and because, also, of the presence of certain organic matters which interfere to some extent. Therefore proceed as follows: Measure out accurately into a platinum or porcelain dish 10 Cc. of the urine and add about 2 Gm. of potassium nitrate and 1 Gm. of dry sodium carbonate, both free from chlorides. Evaporate to dryness on the water-bath and then heat over the free flame, at first gently and finally to a higher temperature until the mass fuses. The organic matter is destroyed by the nitrate leaving finally a white molten residue. Allow it to cool, dissolve in water and add enough pure nitric acid to give a faint acid reaction. This destroys the carbonate. The slight excess of nitric acid must in turn be neutralized and this is done by adding a little precipitated and thoroughly washed (chlorine free) calcium carbonate. Pour the solution now into a flask, rinse the dish thoroughly and add the rinsings to the liquid in the flask. Next add 2 drops of the chromate indicator and then the $\frac{N}{10}$ silver solution, until the faint red of silver chromate mixed with the chloride appears showing the end of the titration. If the urine contains sugar or albumin in more than traces evaporate and heat with the sodium carbonate first and then add the nitrate, a little at a time, to the charred mass to avoid too explosive an oxidation. The titration of the residue from the urine is therefore similar to the process by which the correctness of the silver solution was determined above. If in the titration 22 Cc. of the silver solution were used we have $22 \times 3.537 = 77.81$ Mg. of chlorine in the 10 Cc. of urine, corresponding to 128.4 Mg. of sodium chloride. The sodium carbonate should not be omitted in this method as without it there is danger of loss of chlorine by volatilization. With care it gives excellent results but at present is not as generally employed as is the next one.

Volhard's Method. We have here a method by which the chlorine in urine can be quickly and accurately determined without fusion. The principle involved in the process is this. If to a chloride solution a definite volume of standard silver solution be added, and this in excess of that necessary to precipitate the chloride, the amount of this excess can be found by another reaction, subtracted and leave as the difference the volume actually needed for the chloride. The reaction for the excess depends on these facts. A sulphocyanate solution gives with silver nitrate solution a white precipitate of silver sulphocyanate, AgSCN. It also gives with a ferric solution a deep red color due to the formation of soluble ferric sulphocyanate, $Fe_2S_6(CN)_6$. If the silver and ferric solutions are mixed and the sulphocyanate added the second reaction does not begin until the first is completed, that is, the silver must be first thrown down as white sulphocyanate before a permanent red shade of ferric sulphocyanate appears. The presence of silver chloride interferes but slightly with these reactions. Therefore, if we have a sulphocyanate solution of definite strength we can use it with the ferric indicator to measure the *excess* of silver used after precipitating the chlorine of a solution.

The reaction between silver nitrate and a sulphocyanate is expressed by the following equation:

$$AgNO_3 + NH_4SCN = Ag\ SCN + NH_4NO_3.$$

For 16.954 Mg. of the silver nitrate we use 7.597 Mg. of the sulphocyanate. In this method the standard solutions required are

(*a.*) **Standard Silver Nitrate Solution,** $\frac{N}{10}$. Made as before with 16.954 Gm. of the fused salt to the liter. As the solutions are used with nitric acid present the standard can also be made by weighing out accurately 10.766 Gm. of *pure* silver and dissolving it, in a flask, in pure nitric acid. Most of the excess of nitric acid is removed by evaporation and air is blown through to drive out nitrous fumes. The solution is cooled and diluted to 1 liter.

(*b.*) **Standard Sulphocyanate Solution,** $\frac{N}{10}$. This may be made of the potassium or ammonium salt, but the latter is more commonly used. Weigh out about 7.7 Gm. of the pure crystals, dissolve in water and make up to 1 liter. Determine the exact strength as explained below. The true weight cannot be weighed out directly because the otherwise pure salt is frequently a little moist, and because further a salt, pure to begin with, undergoes frequently a slight change on standing.

(*c.*) **Ferric Solution as Indicator.** Use for this a nearly saturated solution of ammonium ferric sulphate (ferric alum) free from chlorine.

To find the exact strength of the sulphocyanate solution proceed as follows: Measure into a flask or beaker 25 Cc. of the $\frac{N}{10}$ silver solution and add to it 2 or 3 Cc. of the ferric indicator. This gives some color and a slight opalescence. Now add about 2 Cc. of pure strong nitric acid, which removes the color and clears the mixture. Into this, from a burette, let the sulphocyanate solution flow, a little at a time, shaking after each addition. A red color appears temporarily, but vanishes on shaking. After a time this red disappears more slowly, which shows that the end of the reaction is near. The burette solution is therefore added more carefully, best by drops, until at last a single drop is sufficient to give a permanent reddish tinge. Something less than 25 Cc. should be used for this. Repeat the test and if the same result is found dilute the sulphocyanate solution so as to make 25 Cc. of the volume used in the titration. For instance, if 24.2 Cc. were required 900 Cc. of the solution may be diluted in this proportion:

$$24.2 : 25 :: 900 : x, \therefore x = 929.8.$$

We have now a standard sulphocyanate solution corresponding exactly to the silver solution. To test it further and illustrate its use with chlorides measure out 25 Cc. of the $\frac{N}{10}$ sodium chloride solution described a few pages back, and add to it, from a burette, exactly 30 Cc. of the silver nitrate solution, then the ferric indicator and the nitric acid as given above. Shake the mixture and filter it through a small filter into a clean flask or beaker. Wash out the vessel in which the precipitate was made with about 20 Cc. of pure water, pouring the washings through the filter. Then wash the filter with about 20 Cc. more of water allowing the washings to mix with the first filtrate. This mixed filtrate contains all the silver used in excess of the chloride. Now bring it under the sulphocyanate burette and add this solution until a reddish tinge becomes permanent. Exactly 5 Cc. should be necessary for this.

The chlorides of the urine may be treated in about the same manner. To a measured volume of the urine, usually 10 Cc., an excess of silver nitrate solution is added, 25 Cc. with most urines is enough, and then the indicator and acid. The mixture is filtered, the precipitate washed and in the filtrate the excess of silver is found by sulphocyanate as above. But it occasionally happens that the addition of nitric acid to the urine develops a red color which obscures the end reaction. To avoid this the urine titration should always be made in the following manner:

To 10 Cc. of urine add 2 to 3 Cc. of pure strong nitric acid, then the ferric indicator and three drops of a saturated, chlorine free solution of potassium permanganate. This gives at first a very deep red color, but it soon fades and with it the urine colors, by oxidation. It is not well to add more of the permanganate than here given.

To the yellow solution add now 25 Cc. of the standard silver solution, shake well and filter. Wash the beaker and filter thoroughly as above described, and in the mixed filtrate and washings find the excess of silver. If in doing this the first drops of the sulphocyanate solution added produce a deep red color it shows that too little silver had been used in the first place. Make therefore, a new test, using more silver nitrate solution, 30 to 50 Cc.

To illustrate, if we use for 10 Cc. of urine, 25 Cc. of silver nitrate, then 3.4 Cc. of the sulphocyanate, $25-3.4=21.6$, the amount of silver nitrate solution actually needed for the chlorides. $21.6 \times 3.537 = 76.40$ Mg. of chlorine in the 10 Cc. of urine, which corresponds to 126.07 Mg. of NaCl in the 10 Cc. or to 12.607 Gm. per liter. If the urine tested had a specific gravity of 1.020 this would equal 1.24 per cent.

Other Mineral Substances.

Several other groups of compounds occur in the urine which may be briefly referred to. The sulphates of sodium and potassium are found to the extent of about 2 Gm. daily in average urine, and in traces certain ethereal sulphates also occur. The clinical significance of these bodies is not very clear, and besides we have no volumetric method accurate enough for their determination. The gravimetric methods by which they may be measured can not be described in this place.

In normal urine several carbonates are often found. The carbonate of ammonium is probably always present in small amount, but in fermented urine, either before or after voiding, its amount may be very great, imparting a strong alkaline reaction. Nitrates, nitrites, sulphides, bromides, and iodides after taking certain remedies, and several other salts have been found in the urine in small amount, but no great clinical importance attaches to them.

Chapter XVII.

THE SEDIMENT FROM URINE.

URINE is frequently cloudy when passed and on standing deposits a sediment of the substances imparting the cloudiness. Other urines which may appear perfectly clear at first also throw down deposits after a time. This is always the case with urine allowed to stand long enough to undergo alkaline fermentation, when a precipitate of phosphates forms. The deposit is frequently caused by a change of temperature. Warm voided urine holding an excess of urates may be perfectly clear, but becomes cloudy as its temperature goes down with the formation of a light reddish sediment. This is a perfectly normal action, and indeed most sediments may be considered in the same light. Urine containing a deposit is not necessarily pathological.

There are conditions, however, in which the sediment is an indication of abnormality, and its examination becomes important clinically. Certain sediments are pathological because of their origin, others because of their amount. For instance, blood and pus corpuscles, casts of the uriniferous tubules of the kidney and a few other forms are not found normally in urine, and their presence is of importance, whether observed in large or small quantity. Sediments containing phosphates, uric acid and urates, calcium oxalate and other salts are common enough and usually attract no attention, but if the amount of these deposits is very large there may be attached to them clinical significance and they deserve study.

In the examination of a sediment it is necessary to allow the urine to stand long enough to deposit the important forms it may contain, which may require twenty-four hours or more. For the deposition of a sediment the urine should be left in a place with an even temperature, preferably not above 15° C. A low temperature favors the precipitation of urates, while decomposition may begin if the temperature be allowed to go up. Some of the light organic forms have a specific gravity so little above that of the urine that they may remain a long time in suspension. It is important, therefore, to allow plenty of time for these to settle. If the weather is warm and there is no good means at hand for keeping the temperature of the urine down until the examination can be made, or if for any reason this must be delayed for some days, it is well to add some preservative to the urine, *i. e.*, something to prevent fermentation. Many substances have been suggested for this purpose, some of which are very objectionable inasmuch as they form precipitates which often obscure what is sought for. Chloroform is the simplest and at the same time one of the best substances which can be added.

To 100 Cc. of the urine to be set aside for tests add 3 or 4 drops of chloroform and dissolve by shaking. It is not well to add more than this as there is danger of leaving minute droplets undissolved, and these are confusing in the subsequent examination. The chloroform may be applied in the form of aqueous solution. Add about 10 Gm. of chloroform to a liter of distilled water and shake thoroughly; about three-fourths will dissolve at the ordinary temperature; 25 Cc. of this saturated solution may be added to 100 Cc. of the urine to be examined, which is then allowed to stand as before.

After the deposit has settled pour off the supernatant liquid very carefully and by means of a small pipette with a *coarse* opening transfer one or two drops to a *perfectly clean* glass slide. Clean a cover glass with *great care* and by means of small brass forceps lower it on the drops of liquid in such a manner as to exclude air bubbles. This can be done by lowering it inclined to the slide, not paral-

FIG. 46.

Sargent's centrifugal machine.

lel with it, so as to touch the liquid on one side first. In settling down the cover now pushes the air in front of it and gives a field generally free from bubbles. The slide is then examined under a microscope with a magnifying power of 250 to 300 diameters. Either natural or artificial light may be used, but it must not be very bright. A very common mistake in the examination of urinary sediments by the microscope is to employ so high a degree of illumination that the lighter and nearly transparent bodies are completely overlooked.

Recently centrifugal machines have been introduced which may be employed to settle the urine. The latter for this purpose is placed in strong test-tubes which are caused to rotate so rapidly that all suspended matters are thrown to the bottom of the tubes, these being hung by the neck in such a manner that in rotation the bottom flies up and outward. Some of these rotating machines are operated by hand, others by water power or electricity. A very convenient form run by a small water motor is now to be had from dealers. It is simply necessary to attach the motor by means of a rubber tube to a faucet delivering water under ordinary city pressure to obtain all the power necessary.

Where the current furnished for electric lighting by the incandescent system is available, centrifugal machines operated by small electric motors are even more convenient. The cut above shows a small machine which has given excellent results. The wires from the motor in the base of the machine are attached by a socket in place of the common incandescent lamps now everywhere used.

A very high velocity is attained in these machines by aid of which the deposit may be secured in minutes instead of hours. The sediment is left in a very concentrated condition in the bottom of the tube, and from it the supernatant urine can be poured much better than when it precipitates in a beaker or bottle in the usual manner.

Sediments from urine are commonly classed as *organized* and *unorganized*, these divisions being then subdivided according to various plans. The important forms under each division are shown in the following schemes:

Organized Sediments.
- Blood corpuscles.
- Mucus and pus corpuscles.
- Epithelium from various locations.
- Mucin bands, or threads.
- Casts of the uriniferous tubules.
- Spermatozoa.
- Fragments of cancer tissue.
- Fungi.
- Certain other parasites.

Unorganized Sediments.
- Uric acid.
- Various urates.
- Leucin and tyrosin.
- Cystin.
- Cholesterin.
- Fat globules.
- Hippuric acid.
- Calcium carbonate.
- Calcium phosphate.
- Calcium oxalate.
- Magnesium phosphates.

In addition to these there are often found in the urine certain bodies whose presence must be called accidental, for instance hairs, fibers of cotton, silk or wool, starch granules, bits of wood, mineral dust, etc. Some of these will be referred to later.

Organized Sediments.

Blood Corpuscles.

Urine containing blood presents a characteristic appearance easily recognized, unless it be present in very small quantity. If the reaction of the urine is acid the color is generally dark; but if alkaline the shade is inclined to reddish. Blood corpuscles enter the urine from several different sources and their presence is usually a pathological indication, but not always, as they may come, for instance, from menstruation. The kidneys, or their pelves, the ureters, the bladder, the urethra, the vagina or the uterus may be the seat of the lesion from which the blood starts, and its appearance sometimes gives a clue to its origin.

Fresh blood corpuscles are clear in outline and show distinctly their bi-concavity. But corpuscles which have

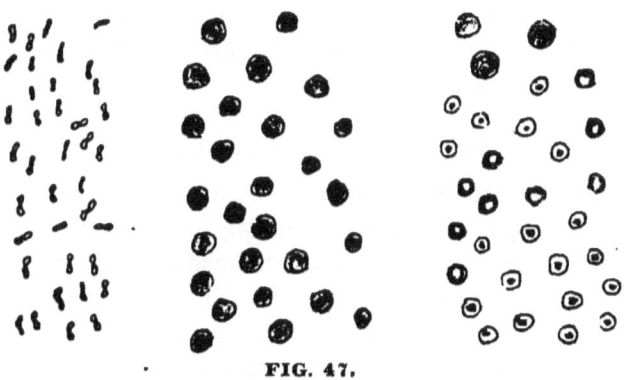

FIG. 47.

Human blood corpuscles, 400 diameters.

been long in contact with the urine become much swollen, less distinct in outline, often biconvex, or nearly spherical even, and lighter in color. As long as the reaction of the urine is acid the corpuscles remain comparatively fresh in appearance but with the beginning of the alkaline reaction disintegration and loss of color soon set in.

The microscopic recognition of blood in urine is easy enough if it is not too old. The fresh, red corpuscles of human blood have a mean diameter of about .0077 Mm., but when swollen by absorption of water they are somewhat larger. When seen on edge they appear as shown at the left in the figure above. If presenting the flat side to the eye they appear as discs whose centers grow alternately light and dark by changing the focus of the instrument. In old urine, especially with alkaline reaction they appear as granulated spheres, shown in the center of the figure. In all cases the color is more or less yellowish. It is generally assumed that the paler washed out corpuscles come from lesions higher up, from the pelvis, or kidney even, while the brighter fresh blood points to a lesion nearer the point of discharge, that is from the bladder or urethra. This is pretty certain to be the case if the blood is discharged but little mixed with the urine and settles rapidly as a distinct mass.

Mucus and Pus Corpuscles.

These are white corpuscles somewhat larger than the red blood corpuscles and spherical in outline. The term *leucocyte* is frequently applied to these as well as to the so-called white corpuscles of blood. Their size varies greatly but the average diameter may be given as .009 Mm. All these corpuscles present when fresh a slightly granular appearance and occasionally show one or more nuclei. The addition of a little acetic acid to the sediment brings the nucleus out distinctly so that it may be seen under the microscope as a characteristic appearance.

Mucus corpuscles in small number are normally present in urine, but pus corpuscles enter the urine as a constituent of pus itself which is an albuminous product discharged from suppurating surfaces and not normal. In a former chapter it was shown that the reactions of mucus and albumin are distinct, but urine containing pus always affords reactions for albumin. Pus in urine tends to form a sediment at the bottom of the containing vessel, and may be recognized by the following method:

Donne's Test. Pour the urine from the sediment and add to the latter an equal volume of thick potassium hydroxide solution, or a small piece of the solid potassa. Stir with a glass rod. The strong alkali converts the pus into a thick viscid mass closely resembling white of egg. Sometimes this is so thick that the test-tube containing it can be inverted without spilling it. In alkaline urine this glairy mass is sometimes spontaneously formed.

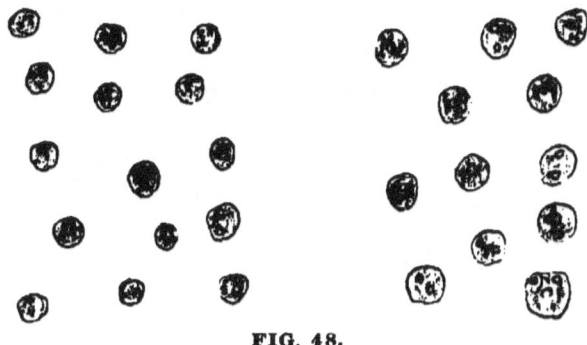

FIG. 48.

Pus corpuscles, 400 diameters.

The appearance of the mucus or pus corpuscles in urine depends largely on the concentration of the latter. In urine of low specific gravity the corpuscles absorb water and swell to larger size than normal, while in a highly concentrated urine they may give out water and become reduced in size and shrunken in appearance.

To recognize them under the microscope transfer a few drops of the sediment to a slide and cover as usual. If the nuclei are not distinct place a drop of diluted acetic acid on the slide at the edge of the cover glass. Part of the acid will flow under the cover and mix with the urine. As it does this the clearing up of the corpuscles with appearance of the nuclei can be very easily followed. Urine

306   THE ANALYSIS OF URINE.

containing much pus is white and milky. The same appearance is often noticed with an excess of earthy phosphates, but the latter clear up with acids while the pus does not.

Epithelium Cells.

Epithelium cells from different sources may appear normally in the urine, and the light cloud which separates from normal urine on standing consists chiefly of these cells. When present in small amount this epithelium has usually no clinical importance, as it easily finds its way into the urine from the bladder, vagina or urethra. An

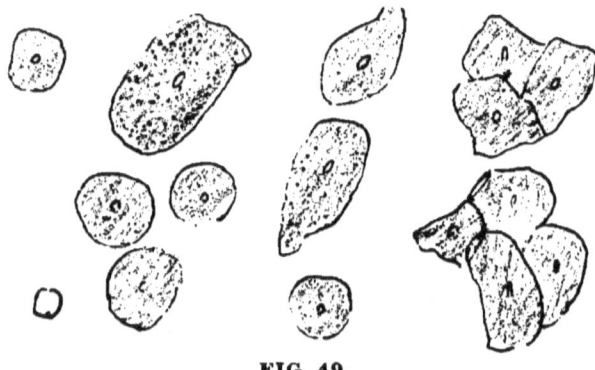

FIG. 49.

Scaly and spherical epithelium.

abundance of cells from these organs would, however, be considered pathological, pointing to a catarrhal condition.

Unfortunately, it is not possible in all cases to determine the source of the cells, as found in urine, partly because cells from different localities have frequently the same general appearance, and partly because, owing to immersion in the urine, they become greatly changed from what they are in the tissue as shown by the microscopic study of sections. It is customary to make three rough divisions of the cells as found in the urine:

1. Spherical cells. 2. Columnar or conical cells. 3. Flat or scaly cells.

The spherical cells are probably normally much flattened but by absorption of water they become swollen and globular. These cells may be derived from several sources, as from the uriniferous tubules or from the deeper layers of the lining membrane of the pelvis of the kidney, or the bladder, or the male urethra.

These cells have a well-defined nucleus resembling that of a pus cell. But they are much larger, and besides show

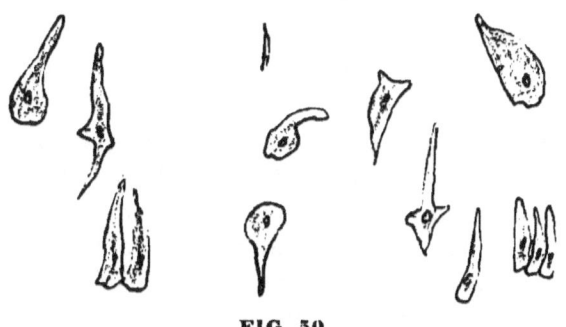

FIG. 50.
Conical epithelium.

the nucleus without addition of acid. In nephritis, or other structural diseases of the kidney, these round cells are found along with albumin, and their recognition is then a matter of importance as indicating a breaking down of the tubular walls. Sometimes these cells form a variety of tube cast, to be described later. But it must be remembered that we cannot distinguish with certainty between the cells from the tubules and those from the other localities mentioned.

Conical cells come generally from the pelvis of the kidney, from the ureters and urethra. Some of these cells are furnished with one or two processes, and are broad in the middle and taper toward each end, while the others are broad at the base and taper to a point.

The large flat cells come from the vagina or bladder, and it is generally impossible to distinguish between them. Sometimes they are very nearly circular, sometimes irregularly polygonal in outline. Sometimes the vaginal epithelium is found in layers of scales, which appear thicker and tougher than the cells from the bladder, which occur singly.

What was said about the decomposition of blood or pus cells in urine obtains also for the various epithelium cells. In acid urine they may maintain their distinct outlines many days, but in alkaline secretion they soon undergo disintegration, which makes their recognition practically impossible. In general the greatest importance attaches to the cells from the tubules of the kidney. The presence of albumin in more than minute traces in the urine would suggest that any smaller spherical cells present may have had their origin in the kidney rather than in the bladder or male urethra. In general it may be said that urine containing large numbers of the smaller, round tubule cells with albumin will also show casts.

Mucin Bands

Urine containing much mucus sometimes exhibits a deposit consisting of long threads or bands, curved and bent in every direction. These bands are important because they are sometimes confounded with the tube casts to be described next. They can be produced in urine highly charged with mucus by the addition of acids, and appear therefore sometimes spontaneously when the urine becomes acid. These threads are sometimes covered with a fine deposit of granular urates and these bear some resemblance to granular casts. In general, however, they are relatively longer and narrower than the true casts of the uriniferous tubules. The mucin threads can occur and frequently do occur in urine entirely free from albumin, while true tube casts are usually associated with albumin, although not always, as will be explained below. The

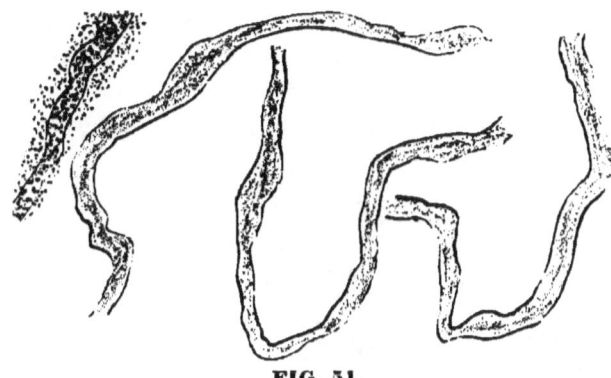

FIG. 51.

Mucin bands or threads.

length and shape of the mucin threads may generally be relied upon to distinguish them from true casts.

Casts.

The structures properly termed casts are seldom found in urine which does not contain albumin. They are formed in the uriniferous tubules, and to a certain extent, are "casts" of portions of the same. Their specific gravity differs but little from that of the urine, for which reason they remain long in suspension. It is therefore necessary to allow the urine to stand some hours at rest, over night or longer, before attempting an examination.

Casts of the uriniferous tubules rarely appear in normal

urine and their recognition is therefore a matter of the highest importance in diagnosis. Much has been written on the subject of the origin of these bodies in the kidney and several theories have been advanced to account for their formation and chemical constitution. Most of this discussion would be out of place in a work like the present dealing mainly with questions of analysis, but enough will be given to aid the student in his examinations. It must be said that few subjects are more perplexing to the beginner than that of their certain recognition, because of the fact that some varieties are so transparent as to be almost invisible, while others are closely resembled by formations

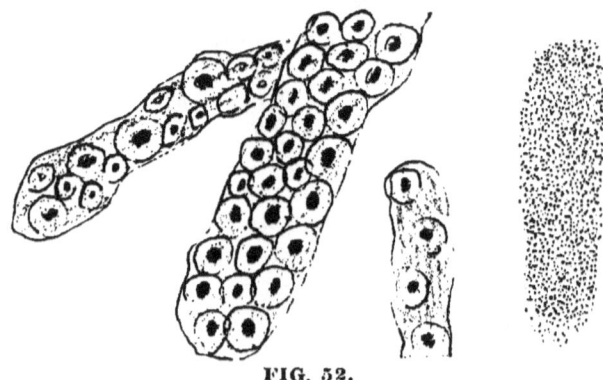

FIG. 52.

Epithelium casts. To the right a bunch of urates or false cast.

of entirely different nature not pathological. With practice, however, these difficulties can be surmounted.

Most of the bodies termed casts are formed of organized structures or the remains of such, but another and rather common form consists of crystalline matter, usually uric acid or fine granular urates.

These bunches of urates have no pathological significance and are of frequent occurrence. Urine containing them clears up by heat and the deposits themselves are dissipated by weak alkali. While it is true that they resemble to some degree the so-called granular casts referred to below there are certain well-defined points of difference. The bunches of urates lack the coherence which can be observed in the true casts, and besides the granulation is finer and more clearly defined.

The fact that mucin bands occasionally appear covered with a precipitate of granular urates has been referred to. These aggregations are more compact than the loose bunches of urates just mentioned and much longer

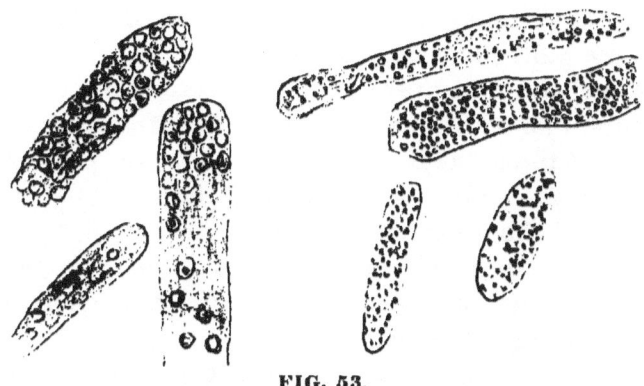

FIG. 53.
Blood casts and granular casts.

generally. They are also darker and therefore more easily seen than are the casts proper or the urates.

The true casts are made up of matter in which evidence of cell structure or transformation is visible. An accurate classification of these bodies cannot yet be made, and, as said, authors differ regarding the importance of several forms and their origin. But for our purpose it will be sufficient to make the following rough division, which accords in the main with what is found in the text-books of urine analysis:

1, Blood casts; 2, Epithelium casts; 3, Granular casts; 4, Fatty casts; 5, Waxy casts; 6, Hyaline casts.

What are termed blood casts consist of or contain coagulated blood recognized by the corpuscles. Plugs of this coagulated matter are forced out from the tubules by pressure from behind, and form one of the most characteristic varieties of casts. They are generally very dark in color, and easily distinguished from other matter. A representation of blood casts is given in the preceding cut.

In epithelium casts the characteristic substance is the lining epithelium of the tubule. Sometimes this lining epithelium becomes detached in the form of a hollow cylinder, the walls consisting of the united cells. Again, the coagulated contents of the tubule in passing out may carry the epithelium with it as a coating. In either case a grave disorder of the kidneys is indicated, as acute nephritis, or other disease in which a profound alteration of the internal structure of the organ is involved.

What are termed granular casts, proper, appear in a variety of forms, produced probably by the disintegration of blood or epithelium casts.

There is no uniformity in the fineness of the granulation; sometimes a high amplification is necessary to disclose the structure. Occasionally blood corpuscles, epithelium, fat globules and crystals can be detected in them, and when derived from blood cast disintegration they usually have a yellowish red color, which makes their recognition comparatively easy. In outline they are generally regular, with rounded ends, one of which is somewhat pointed. Frequently, however, they appear to be broken, the ends showing irregular fracture.

Fatty casts contain oil drops produced by some variety of fatty degeneration of the tissues of the kidney. These oil drops may form coherent bunches, or they may be held by patches of epithelium. It also happens that epithelium or granular casts may be partially covered by oil drops. The name *fatty cast* is applied to those in which the fat globules predominate. Along with these globules the microscope sometimes shows crystals of free fatty acids, and probably also of soaps containing calcium and magnesium.

Waxy casts consist of the peculiar matter produced by amyloid degeneration of the kidney. They have a glistening waxlike or vitreous appearance, and refract light very

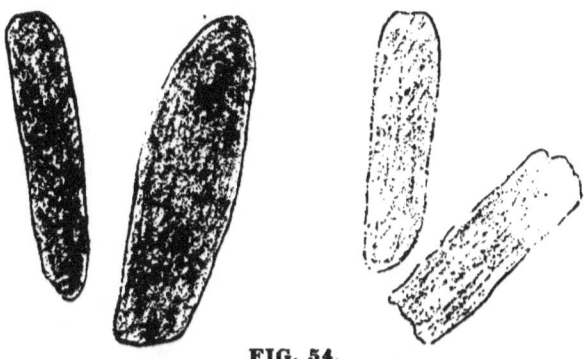

FIG. 54.
Waxy and hyaline casts.

strongly. Sometimes they reach a great length, and they frequently are found with blood corpuscles or oil drops on the surface. They have been detected in several renal disorders. Illustrations are given to the left above.

True hyaline casts are nearly transparent and hard to see unless the illumination is very carefully managed. To detect them it is often necessary to add a few drops of a dilute solution of iodine in potassium iodide to the sediment. This imparts a slight color which renders them visible.

The hyaline casts seem to be formed by the passage of

homogeneous matter from the tubules, leaving the epithelium behind. A cast is rarely perfectly hyaline, as at least an occasional blood corpuscle, fat globule or epithelium cell will usually be found attached to it. Waxy casts may be looked upon as a special form of hyaline casts. Very imperfect representations are given in the above cut.

In general, it must be said that the representation of these casts on paper is a very difficult matter. Ordinarily they are drawn and printed much too heavy and dark.

Hyaline casts do not necessarily indicate kidney disease, although this is usually the case. They have been found in urine free from albumin and under circumstances not connected with renal disorders.

The preservation of sediments containing casts is unusually difficult because of the nature of the material to be preserved. In urine of the slightest alkalinity their disintegration soon begins so that the outlines are rendered indistinct, often making identification impossible.

For temporary preservation the addition of chloroform renders as good service as anything else. Many other sediments can be permanently mounted and kept for future comparison but with casts this can rarely be done.

Beginners are apt to overlook casts in their first examinations. It must be remembered that some of them are nearly transparent and unless brought into proper focus they may not be seen at all. At the outset students usually employ too bright a light in looking for casts. While no specific directions can be given regarding the intensity of illumination best suited to the purpose, this may be said that the light. commonly found necessary in studying ordinary histological slides is far too bright to use in the search for casts.

Practice alone, first under the direction of the instructor, will indicate what is proper here.

SPERMATOZOA.

These minute bodies, as found in the semen of man, have a mean length of about .050 Mm. Nearly one-tenth of this is in the head portion. When observed in recently discharged semen they have a characteristic spontaneous movement by which they are propelled forward rapidly. This motion is soon lost if the semen is diluted with water or similar liquid. Hence, as usually seen in urine they are entirely motionless. They are found abundantly in the urine of men after coitus or nocturnal emissions, and also

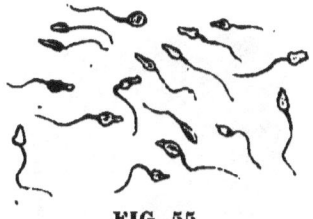

FIG. 55.

Human spermatozoa.

in spermatorrhœa, when their presence is continuous and characteristic.

In the urine of women they are likewise found after connection, and their detection here has often interest from the medico-legal standpoint. The proof of rape can often be established by the recognition of spermatozoa.

Although their motion is soon lost in foreign liquids the substance itself of the spermatozoa is not readily destroyed. In this respect they are more permanent, probably, than all other organized structures found in the urine and can be readily distinguished after many days or months even in urine, or in sediments which have become dry.

For their recognition a power of 250 diameters is sufficient, but 350 to 400 diameters is preferable. With this higher amplification it is possible to readily distinguish between spermatozoa and certain fungus growths bearing some resemblance to them.

Cancer Tissue.

Fragments of cancerous and other morbid growths are occasionally met with in the urine, and when certainly recognized are an important aid in diagnosis.

The recognition of cancer cells along with the several kinds of epithelium cells found in urine is difficult and lies somewhat beyond the usual line of urine examinations.

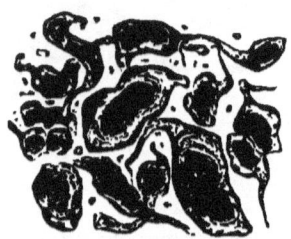

FIG. 56.

Cancer cells which have been seen in urine.

Sometimes the identification is comparatively simple because of the unusual shape of the cells, or when they are present in great number, but this is not often the case.

The cut shows some of these cells figured by Beale.

Fungi.

The urine sometimes contains certain fungus growths the recognition of which is important. These may have entered the urine after voiding or they may have come from the bladder.

Normal urine when passed is probably free from fungi of all kinds, but in a short time certain organisms enter it from the air or from other sources and become active in producing in it characteristic changes.

The three important groups of fungi, the *schizomycetes* or *bacteria*, the *hyphomycetes* or *moulds* and the *blastomycetes* or *yeasts* are represented in the organisms sometimes found in the urine. The conditions under which they are found will be briefly explained.

Of the bacteria the following have been observed:

Micrococcus Ureæ. This is the exceedingly common form found on urine undergoing alkaline fermentation by which urea is converted into ammonium carbonate. It is usually introduced from the air and multiplies very rapidly under ordinary conditions. Nearly all old specimens of urine, unless containing some active preservative are found infected with this small organism. The micrococci are minute spherical bodies belonging to the suborder *sphærobacteria*, and are found separate or in chains. They are the smallest of the organized forms occurring in urine and appear under a power of 250 diameters but little more than points. While generally finding their way into urine after it has been voided they are occasionally present in the bladder. It is usually held that under such circumstances they have been introduced by a dirty catheter or sound, although cases are on record where this has not been proven.

In the bladder they give rise to alkaline fermentation, so that the voided urine may show ammonium carbonate directly.

Streptococcus Pyogenes is a pathogenic form sometimes found in the urine in cases of infectious diseases.

Sarcinæ. The genus *sarcina* is frequently classed with the sphærobacteria and several species have been found in urine. The cells are larger than those of micrococcus ureæ, and are arranged in groups of two or four usually. They are not pathogenic.

Bacilli. Several species of the genus *bacillus* are found in urine in disease. The most important of these are the typhoid bacillus, *bacillus typhi abdominalis*, the tubercle bacillus, *bacillus tuberculosis* and the bacillus of glanders, *bacillus mallei*. These bacilli occur in urine only during the progress of the corresponding diseases and their detection is of the highest interest. A description of the

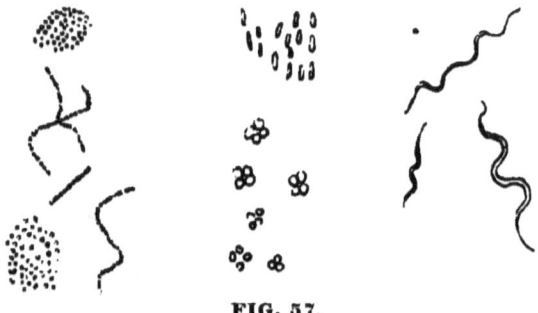

FIG. 57.

Micrococci and other bacteria.

methods to be followed for the certain demonstration of these bodies is not within the scope of this book, but must be looked for in the laboratory manuals of bacteriology.

Spirilla. Certain species of the genus *spirillum* have been found in urine. The best known of these is the spirillum of relapsing fever, *spirillum Obermeieri*. This is only found rarely and as its habitat is the blood of relapsing fever patients it must enter the urine through a hæmorrhage into the kidney. Its form is that of a long, wavy spiral, which makes its detection somewhat easy.

Although not pathogenic it is well to call attention to certain moulds which may sometimes be seen in urine. The common blue-green mould, *penicillium glaucum*, is the best known of these, and is occasionally found in urine along with yeast cells. Another mould which has been found in urine is the *oidium lactis*, commonly occurring in milk and butter. It has been observed in fermenting diabetic urine. Both of these fungi enter the urine after voiding. In urine which has stood sometime in a cool place the *penicillium glaucum* sometimes becomes covered with an incrustation of urates or minute crystals of uric acid.

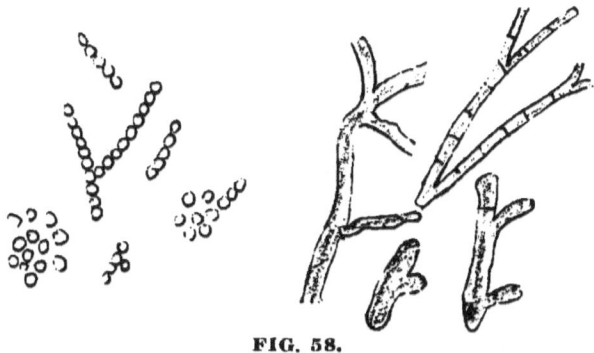

FIG. 58.

Yeast cells and common mould.

Finally we have yeast cells in urine and sometimes in great numbers. Like other fungi they enter the urine from the air and when not very abundant have no significance. In great numbers the yeast cells suggest presence of sugar. The ordinary yeast plant, *saccharomyces cerevisiæ*, is shown, isolated and budding, in the accompanying figure.

OTHER PARASITES.

Besides fungi the urine in rare cases contains other parasites of animal as well as vegetable nature. Some of these enter it accidentally from the air and have no interest, but a few are present in the urine when passed, and are of importance. Among these are certain thread worms, the eggs of worms and hooklets of a species of tapeworm, the *taenia echinococcus*.

These forms can appear in the urine only through rupture from some other organ, and while rare here are common enough in Egypt and other tropical countries. Urine containing eggs of worms, the worms themselves or fragments usually contains blood or other evidence of rupture.

Chapter XVIII.

UNORGANIZED SEDIMENTS AND CALCULI.

Unorganized Sediments.

Uric Acid.

AMONG the more common of the unorganized sediments found in urine this must be mentioned first. As was explained in the last chapter uric acid occurs normally in combination in all human urine.

Some time after its passage urine often undergoes what has been spoken of as the *acid fermentation* by which a precipitate of urates and even free uric acid may appear. This reaction is in no sense due to a ferment process in the ordinary sense of the term, but is probably brought about by a purely chemical double decomposition. Urine contains acid sodium phosphate and neutral sodium urate and it has been suggested that these react on each other according to the following equation:

$$Na_2C_5H_2N_4O_3 + NaH_2PO_4 = NaC_5H_3N_4O_3 + Na_2HPO_4$$

The precipitate of acid urate settles out and forms a light reddish deposit. If the amount of acid phosphates present is excessive the reaction may go still further, resulting in the precipitation of free uric acid. The well characterized crystals of uric acid are often found with the sediment of fine urates. Sometimes this liberation and precipitation of the acid takes place in the bladder and the urine as passed shows the crystals or "gravel." If they are relatively large, which is sometimes the case, their passage through the urethra may cause severe pain.

As the illustrations show uric acid occurs in a great variety of forms. The rosettes and whetstone shaped crys-

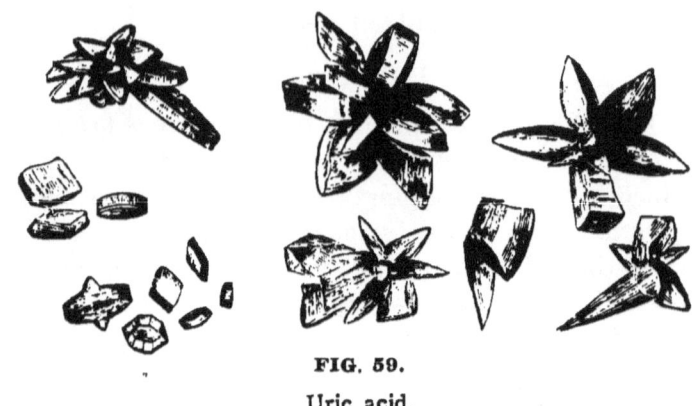

FIG. 59.
Uric acid.

tals are probably the most common, while long spiculated forms are frequently seen. Pure uric acid is colorless but

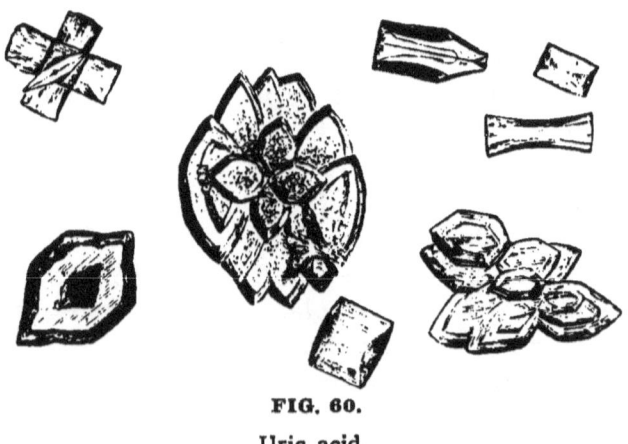

FIG. 60.
Uric acid.

as deposited from urine it is always reddish yellow, because of its property of carrying down coloring matters.

The crystals are often so large that their general form can be seen by the naked eye; usually, however, they are minute.

Uric acid crystals when once deposited are not readily redissolved by heat, but they go into solution by the addition of alkali. If the urine contains extraneous matter, as specks of dusts, bits of hair, cotton or wool fibers the crystals are very apt to deposit on them.

URATES.

The common fine sediments of urine are usually urates or amorphous phosphates. They can be most readily dis-

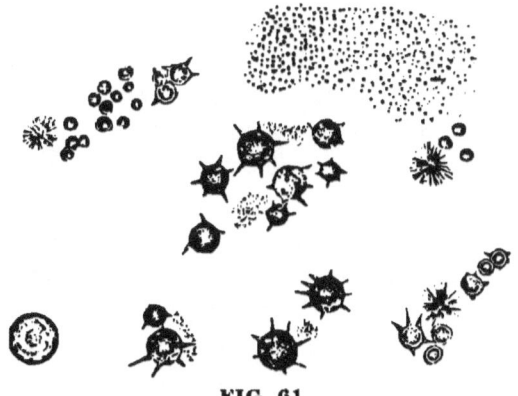

FIG. 61.

Common crystalline and granular urates.

tinguished by their behavior with acids and on application of heat. Urates disappear on warming the urine containing them, while a phosphate sediment is rendered more abundant. A urate sediment is little changed by acids, while the phosphates dissolve completely if the urine is made acid in reaction with hydrochloric or nitric acid. The acid urates of sodium and ammonium are the must abundant and are shown in the cut. Acid ammonium urate may exist in urine which has become alkaline from the decomposition of urea and formation of ammonium carbonate, and may therefore be seen in company with the phosphate sediments. The other urates dissolve in alkaline urine. Like uric acid the urates appear in a great variety of forms, and there is still some uncertainty about the composition of some of their crystals which have been found in urine.

Leucin and Tyrosin.

These two substances are of rare occurrence in urine and appear only under pathological conditions. Urine containing them shows usually strong indications of the presence of biliary matters as they generally are found in consequence of some grave disorder of the liver in which destruction of its tissue is involved. They have been most frequently found, and associated, in acute yellow atrophy of the liver and in severe cases of phosphorus poisoning. In general they must be considered as products of disintegration and are produced in the intestine in large quantity by bacterial agency in the last stages of the digestion of albuminoids, as was pointed out in an earlier chapter.

As both bodies are slightly soluble they may not be seen directly, but only after partial concentration of the urine. In pure condition leucin crystallizes in thin plates but from urine it separates in spherical bunches made up of fine plates or needles. These bunches are sometimes so compact that it is hard to distinguish between them and other substances, particularly lime soaps and oil drops. Chemical tests must therefore be applied. If mercurous nitrate is added to a leucin solution and the mixture is warmed metallic mercury precipitates. This test can be carried out only when the substance is abundant enough to be purified by crystallization from hot water. Pure leucin when strongly heated with nitric acid on platinum forms a colorless residue, which when heated with potassium hydroxide leaves an oil-like drop that does not wet the platinum.

Tyrosin is usually seen in long needles, which sometimes are bunched in the form of sheaves and more readily recognized than is leucin. Tyrosin heated with nitric acid on platinum turns orange yellow, and leaves a dark residue which becomes reddish yellow by addition of caustic alkali. Solutions containing tyrosin when treated

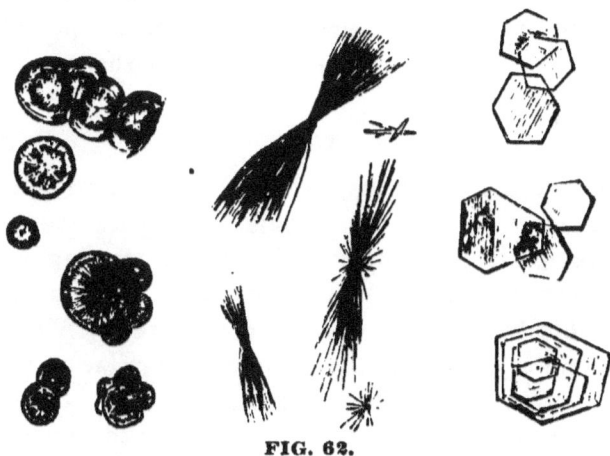

FIG. 62.

Leucin spheres, tyrosin needles and cystin plates.

hot with mercuric nitrate and potassium nitrite turn red and finally throw down a red precipitate.

Leucin and tyrosin may be present in the urine yet not show as a sediment. For their detection under these circumstances precipitate the urine with basic lead acetate and filter. Separate the excess of lead from the filtrate by hydrogen sulphide and filter again. Concentrate the filtrate on a water-bath to a syrup and treat it with a little *absolute* alcohol to remove urea. Some leucin may dissolve with the urea in this treatment. Next boil the residue with 60 to 70 per cent alcohol, filter, concentrate the filtrate to a small volume and allow it to stand in a cool place for crystallization. If crystals appear their form indicates whether they consist of pure tyrosin or a mixture of tyrosin and leucin. The latter, being more soluble in strong alcohol, can be separated by washing with this liquid. The final tyrosin residue can be used for the tests given above.

The alcoholic solutions may contain leucin, which can be recognized after evaporation. Both leucin and tyrosin decompose readily in urine undergoing putrefactive changes; it is therefore necessary to apply the test to urine as fresh as possible.

Cystin.

This is a rare sediment, although it is found constantly in the urine of certain individuals. It crystallizes in thin hexagonal plates, small ones sometimes resting upon or overlapping large ones. The crystals are regular in form but variable in size and readily recognized. A rare form of uric acid crystallizes in a somewhat similar manner but the two substances differ in their behavior toward ammonia. To distinguish between them in the microscopic test place a drop of ammonia water on the slide and allow it to pass under the cover glass. Cystin dissolves, but, unless heated, uric acid does not. When the ammonia evaporates cystin reprecipitates.

Cystin is precipitated from urine by addition of acetic acid. Mucin and uric acid may come down at the same time. The precipitate is collected on a filter, washed with water and finally dissolved in ammonia. By neutralizing the ammoniacal filtrate with acetic acid and concentrating a little, it comes down in the characteristic form suitable for microscopic recognition.

CHOLESTERIN.

This substance occurs occasionally in urine, and possibly only in cases of cystitis. It is recognized by its characteristic crystalline form, large, thin plates, shown in the following cut. These plates are nearly transparent but from their size cannot be overlooked.

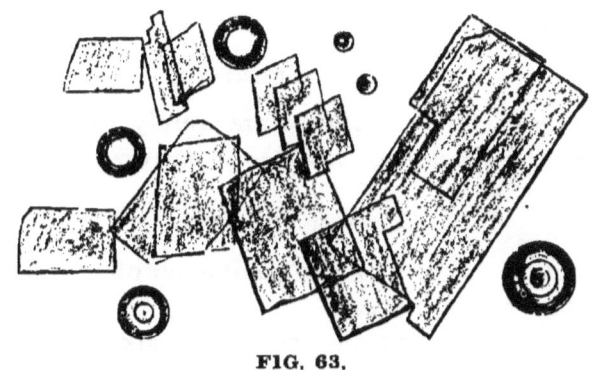

FIG. 63.

Cholesterin plates and fat globules.

FAT GLOBULES.

These are often seen in urine, but in most cases have not been voided with it. They can come from several extraneous sources, as from a catheter, from vessels in which the urine is collected or sent for examination, from admixed sputum, etc., which facts should be borne in mind.

Fat has been found in cases of fatty degeneration of the kidney and more abundantly in chyluria where communication seems to be formed between the lymphatics and the urinary tract by the invasion of the small thread worms referred to above.

HIPPURIC ACID.

This acid is found normally in human urine in small amount. It may be found in large quantity after taking benzoic acid and may even appear in crystalline form in the sediment. It has no pathological importance, ordinarily.

Calcium Carbonate.

This is sometimes observed as a coarse, granular sediment which dissolves with effervescence in acetic acid. It occasionally forms dumb-bell crystals, and is devoid of pathological importance.

Calcium Sulphate.

Crystals of this substance are rarely found in urine. They form long colorless needles or narrow, thin plates.

Calcium Oxalate.

We have here one of the commoner of the crystalline bodies observed in urine.

This may be found in neutral or alkaline urine, but more commonly in that of acid reaction. It occurs normally and sometimes is very abundant, especially after the consumption of vegetables containing oxalic acid.

Two principal forms of the crystals are found, the octahedral and dumb-bell crystals.

The octahedra have one very short axis which gives the crystals a flat appearance. When seen with the short axis perpendicular to the plane of the cover glass, which is the common position, they appear as squares crossed by two bright lines. Sometimes they are seen on edge, and then present a rhomb in section with one diameter very much shorter than the other.

A form of triple phosphate bears a slight resemblance to calcium oxalate, but it is soluble in acetic acid, while the oxalate is not.

The dumb-bells are much less common than the octahedra, and are found in several modified forms, as shown in one of the figures.

The clinical significance of the oxalate is not clearly understood. It does not seem to be characteristic of any disease even when occurring in quantity. It has been found considerably increased in dyspeptic conditions, but not always, and many of the statements found concerning its significance seem to have been based on insufficient observations.

Urine may contain a large amount of oxalic acid which does not show as a sediment, but must be found by pre-

FIG. 64.

Calcium oxalate.

cipitation by calcium chloride in presence of ammonium hydroxide. Acetic acid is then added in very slight excess and the mixture is allowed to stand for precipitation.

The constant or prolonged excretion of large amounts of oxalic acid is spoken of as oxaluria.

The Phosphates.

It was explained in Chapter XVI. that phosphates of alkali and alkali-earth metals occur normally in the urine, and a method was given for their measurement. As sediment we know several forms of calcium and magnesium phosphates and the microscopic detection of these will be here explained. In normal fresh urine of acid reaction these phosphates are held in solution, but if the urine as passed is alkaline it is often turbid from the presence of basic phosphates held in suspension. Urine which has stood long enough to undergo the alkaline fermentation always contains phosphates in the sediment. Finally, it must be remembered that a neutral or very slightly acid urine, con-

FIG. 65.
Triple phosphate.

taining ammonium salts in abundance, may also deposit a crystalline precipitate of ammonium magnesium phosphate. The common phosphate sediments are those consisting of ammonium magnesium phosphate (triple phosphate), basic magnesium phosphate, neutral calcium phosphate and mixed amorphous phosphates of calcium and magnesium.

Triple Phosphate. Of the crystalline phosphate deposits this is the most abundant and at the same time the most characteristic.

The crystals are the largest found in urine, and from their shape are sometimes spoken of as coffin lid crystals. Ordinarily they are not found in perfectly fresh urine, but after it has undergone the alkaline fermentation they are generally present in profusion.

Basic Magnesium Phosphates. Crystals having the composition $Mg_3(PO_4)_2 \cdot 22H_2O$ are sometimes found in urine of nearly neutral reaction. They consist of thin,

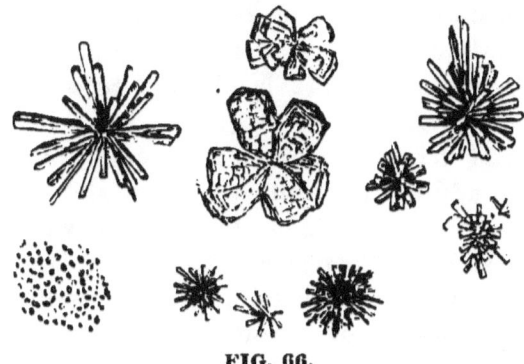

FIG. 66.

Neutral calcium phosphate and amorphous phosphate.

transparent rhombic plates with angles of approximately 60° and 120°. If urine containing this sediment becomes alkaline triple phosphate forms.

Neutral Calcium Phosphate. This has the composition $CaHPO_4 \cdot 2H_2O$ and is found in urine of neutral or slightly acid reaction. It crystallizes frequently in rosettes formed of wedge-shaped single crystals uniting at their apices. The cut above shows some variations in the form.

Amorphous Phosphates. Finally we have the very common, finely granular earthy phosphates in amorphous condition. This sediment dissolves readily in weak acetic acid and is colorless. The common amorphous urate sediment is colored and does not dissolve in acetic acid. On addition of sodium carbonate or hydroxide to urine the precipitate which forms consists mainly of this phosphate.

These several phosphates can be produced artificially and should be made for study and comparison. The neutral (basic) magnesium phosphate can be made by dissolving 15 Gm. of crystallized common sodium phosphate in 200 Cc. of water and mixing this with 3.7 Gm. of crystallized magnesium sulphate in 2000 Cc. of water. Enough sodium bicarbonate is added to give an amphoteric reaction and then the mixture is allowed to stand a day or more for precipitation.

Crystals of triple phosphate of peculiar form are often obtained by adding ammonia to urine, and sometimes a trace of ammonia is sufficient to throw down the crystals of neutral calcium phosphate. The latter can also be obtained by adding to a weak solution of crystallized sodium phosphate a trace of acid and then a very little calcium chloride solution.

Foreign Matters.

The sediment of urine often contains foreign substances which have become mixed with it accidentally. The most common of these are hairs, woolen, cotton or silk fibers, granules of starch, fat globules, dust and sand granules, bits of woody fiber and remains of articles of food. Some of these are represented in the cuts on the following pages.

Urinary Calculi.

Calculi like the sediments just described are formed by the precipitation of certain substances from the urine, but in compact form. Occasionally a calculus consists of a single substance, as calcium oxalate or cystin, but in the great majority of cases a mixture of bodies is present, these being deposited usually in layers around a nucleus which serves as the foundation of the concretion. Calculi are built up much as certain forms of crystals are by successive depositions on a nucleus. Uric acid is a very common nucleus on which may be deposited urates, phosphates, organic matters, etc.

Calculi are sometimes distinguished as *primary* or *secondary*. Primary calculi may be traced to an alteration of

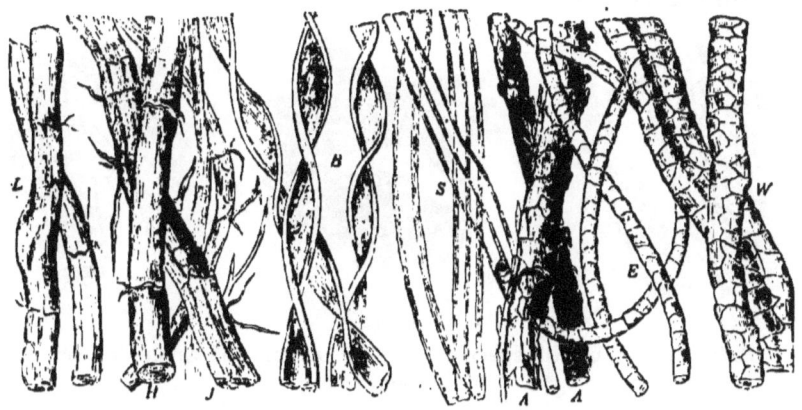

FIG. 67.

L, linen fibers; *H*, hemp fibers; *J*, jute fibers; *B*, cotton fibers; *S*, silk fibers; *A*, alpaca wool; *E*, fine wool; *W*, common wool.

the urine of such a nature that its reaction is constantly acid. The foundation for the concretions in this case is found in the kidney and they are built up of such substances as most easily deposit from acid urine. Secondary calculi are generally formed in the bladder and have as nuclei matters precipitated from alkaline urine, coagulated blood or other organic substances. Sometimes fragments introduced into the bladder from without serve as the foundation for these secondary formations. Bits of catheters, remains of bougies, and other things have been found as the nuclei around which concretions have formed. The recognition of the nucleus is a matter of the first importance as this gives a clew to the determining cause active in the formation of the calculus.

In making an examination, then, of a calculus, it is first cut in two by means of a very sharp thin saw. This exposes the nucleus which may often be recognized by the eye alone. If one of the halves be polished it is often possible to discern distinctly the various layers grouped around the center.

FIG. 68.
Left, pubic hair with spermatozoa; center, hair of woman's head; right, cat's fur.

In a large number of cases examined by *Ultzmann* about 80 per cent were found to contain uric acid as the nucleus.

CHEMICAL EXAMINATION.

In the chemical examination of a calculus several methods may be employed. We may begin by applying certain preliminary tests designed to show the general nature of the stone.

Heat Test. Reduce some of the calculus to a powder and heat to bright redness on platinum foil. Two cases may arise: (a), the powder is completely consumed; (b), the powder is only partially consumed or not at all.

Case (a.) If this is the result of the incineration the following substances may be suspected:

Uric Acid, which may be recognized by dissolving a little of the powder in weak alkali, precipitating by hydrochloric acid and examining the precipitate by the microscope.

Ammonium Urate. This gives the above reaction under the microscope, and is further recognized by the liberation of ammonia when heated with a little pure sodium hydroxide solution.

Cystin. Dissolve some of the powder in ammonia, filter if necessary and allow drops of the filtrate to evaporate spontaneously on a slide. Cystin is then recognized by the microscope as already explained. Cystin contains sulphur which on burning on the platinum foil gives rise to a disagreeable sharp odor. If a little of the powder be heated with a mixture of potassium nitrate and sodium carbonate the sulphur is oxidized to sulphate, which may be recognized by the usual tests.

Xanthin. This is a rare substance in calculi. Those consisting wholly of xanthin are brown in color and take a wax-like polish. A method of recognition will be given below.

Organized Matter. Parts of blood cells, epithelium, precipitated mucin, pus corpuscles and similar substances may become entangled with the growing stone and even form a large part of it. On burning, these bodies are recognized by the characteristic odor of nitrogenous matter.

Case (b.) When an incombustible residue is left on the platinum foil the stone may contain the following constituents:

Calcium Oxalate. Stones of this substance are very hard and break with a crystalline fracture. They are often called "mulberry calculi." When the powder is heated it decomposes, leaving carbonate, which may be recognized by its effervescence with acids.

Calcium and Magnesium Phosphates. These leave a residue in which the metals and phosphoric acid may be detected by simple tests of qualitative analysis. The ignited powder is soluble in hydrochloric acid without effervescence. When ammonia is added to this solution in quantity sufficient to give an alkaline reaction a precipitate of triple phosphate or calcium phosphate appears, which may be recognized by the microscope.

The above tests are generally sufficient to tell all that is practically necessary about the calculus. If more detailed information is desired a systematic analysis may be made according to the following scheme.

Systematic Analysis. 1. Reduce the calculus to a fine powder and pour over it some water and finally dilute hydrochloric acid in a beaker. Warm gently half an hour, or longer, on the water-bath. Then allow to cool and filter.

2. Treatment of the residue. It seldom happens that the calculus is completely soluble in the weak acid. A residue usually remains which may contain uric acid, xanthin, calcium sulphate, and remains of organized matter. To prove the xanthin treat the residue with warm *dilute* ammonia and filter. The filtrate contains the xanthin if it is present. Acidify it with nitric acid and add a small amount of silver nitrate solution. This produces a flocculent precipitate which dissolves by warming, and crystallizes on cooling in bunches of fine needles.

In the residue free from the xanthin look for calcium

sulphate by extracting with water and applying the usual tests. This solution may contain uric acid which is recognized by evaporation and crystallization after adding a little hydrochloric acid. In the final residue some uric acid may be also present. Dissolve in alkali, reprecipitate with hydrochloric acid, and examine any crystals which may form under the microscope.

3. Treatment of the hydrochloric acid solution. This may contain calcium oxalate, cystin, the phosphates, and possibly some xanthin. Look for the last in a small portion of the solution. Make this portion alkaline with ammonia, add a few drops of calcium chloride solution, filter if a precipitate forms and treat the filtrate with ammoniacal silver nitrate solution. In presence of xanthin a flocculent precipitate forms.

Dilute the remaining and larger portion of the hydrochloric acid solution with twice its volume of water, add enough ammonia to give a strong alkaline reaction and then acetic acid to restore a weak acid reaction. By this treatment phosphates are held in solution while calcium oxalate, if present, precipitates. Therefore allow the mixture to stand half an hour and then filter off any precipitate which appears. This precipitate may contain cystin as well as calcium oxalate. Cystin may be dissolved by pouring ammonia on the filter, and on evaporating the ammoniacal solution is obtained in form suitable for microscopic examination.

The residue free from cystin is dried and heated to redness on platinum foil. This treatment converts calcium oxalate into carbonate. Place the foil in a beaker and add some dilute acetic acid; an effervescence shows the carbonate. To the clear solution add next some ammonium oxalate which gives a white precipitate of calcium oxalate, if the latter metal is present.

We have next to look for the phosphates and bases in the acetic acid solution obtained after filtering off cystin and calcium oxalate. More calcium may be present, in excess of that combined as oxalate, which may be recognized by adding a little solution of ammonium oxalate. If a precipitate forms treat the whole of the liquid with the ammonium oxalate, after warming on the water-bath, allow to stand an hour and filter. Concentrate the filtrate in platinum to a small volume, transfer to a large test-tube and add enough ammonia to produce an alkaline reaction. If a precipitate appears now it must consist of magnesium phosphate, showing both magnesium and phosphoric acid as present in the original. If no precipitate appears magnesium is absent but phosphoric acid may still be present. To find it divide the ammoniacal liquid into two portions. To one add a few drops of magnesia mixture, and to the other add nitric acid in slight excess and then ammonium molybdate reagent. Both tests should yield the reactions characteristic of phosphates, if present.

This procedure serves for the recognition of the important constituents of calculi. But ammonium, potassium and sodium compounds are sometimes present and may be recognized readily. To detect ammonium salts the original calculus powder may be heated with pure potassium hydroxide solution, or the hydrochloric acid solution of the calculus may be neutralized and heated with the same solution. The ammonia is recognized by the odor or by its reaction on moistened red litmus paper.

To recognize the alkali metals a solution of the powdered calculus in hydrochloric acid is treated with pure ammonia and a little ammonium carbonate in excess. The precipitate formed is allowed to settle and filtered off. The filtrate is then evaporated to dryness in a platinum dish and the residue strongly heated to drive off all ammonium salts. What is now left contains sodium and potassium if they were present in the original. Moisten this final residue with water and a drop of hydrochloric acid and test with a platinum wire in the flame of a Bunsen burner, using a deep blue glass when looking for potassium. Only a very intense yellow color can be taken as indicative of sodium.

APPENDIX.

Test Solutions and Tables.

TABLE OF ATOMIC WEIGHTS.

(ACCORDING TO MEYER AND SEUBERT.)

Name.	Symbol.	Atomic Weight.	Name.	Symbol.	Atomic Weight.
Aluminum	Al	27.04	Molybdenum	Mo	95.9
Antimony	Sb	119.6	Nickel	Ni	58.6
Arsenic	As	74.9	Nitrogen	N	14.01
Barium	Ba	136.9	Osmium	Os	190.8
Beryllium	Be	9.03	Oxygen	O	15.96
Bismuth	Bi	208.9	Palladium	Pd	106.35
Boron	B	10.9	Phosphorus	P	30.96
Bromine	Br	79.76	Platinum	Pt	194.3
Cadmium	Cd	111.5	Potassium	K	39.03
Cæsium	Cs	132.7	Rhodium	Rh	102.9
Calcium	Ca	39.91	Rubidium	Rb	85.2
Carbon	C	11.97	Ruthenium	Ru	101.4
Cerium	Ce	139.9	Samarium	Sm	149.62
Chlorine	Cl	35.37	Scandium	Sc	43.97
Chromium	Cr	52.0	Selenium	Se	78.87
Cobalt	Co	58.6	Silicon	Si	28.3
Columbium	Cb	93.7	Silver	Ag	107.66
Copper	Cu	63.18	Sodium	Na	23.0
Didymium	Di	142.0	Strontium	Sr	87.3
Erbium	Er	166.0	Sulphur	S	31.98
Fluorine	F	19.0	Tantalum	Ta	182.0
Gallium	Ga	69.9	Tellurium	Te	125.0
Germanium	Ge	72.3	Terbium	Tb	159.1
Gold	Au	196.7	Thallium	Tl	203.7
Hydrogen	H	1.0	Thorium	Th	231.9
Indium	In	113.6	Tin	Sn	118.8
Iodine	I	126.53	Titanium	Ti	48.0
Iridium	Ir	192.5	Tungsten	W	183.6
Iron	Fe	55.88	Uranium	U	239.0
Lanthanum	La	138.2	Vanadium	V	51.1
Lead	Pb	206.4	Ytterbium	Yb	172.6
Lithium	Li	7.01	Yttrium	Yt	88.9
Magnesium	Mg	24.3	Zinc	Zn	65.1
Manganese	Mn	54.8	Zirconium	Zr	90.4
Mercury	Hg	199.8			

List of General Reagents and Test Solutions.

Acid, sulphuric (strong). The commercial acid is sufficient for most purposes, where a strong acid is called for. Where the pure acid is required it is mentioned in the text.

Acid, sulphuric (dilute). Add one part of the above acid to four parts of distilled water, mix thoroughly, and allow to stand twenty-four hours. Then siphon, or pour off the clear liquid, which is ready for use. The strong acid must be poured into the water very slowly, and with constant stirring, to avoid a too sudden elevation of temperature.

Acid, nitric (strong). The strong commercial acid can be employed in most cases where this acid is called for. A pure strong acid is also employed, occasionally.

Acid, nitric (dilute). Where this acid is called for as a reagent it should be made by mixing one part of the *pure* strong acid with four parts of distilled water It should be free from traces of chlorine and sulphates.

Acid, hydrochloric (strong). The yellow commercial acid is largely used in the laboratory in the preparation of other substances It is seldom pure enough to be employed as a test reagent.

A colorless acid, free from organic matter, iron and traces of sulphates, must be used when the pure strong acid is called for.

Acid, hydrochloric (dilute.) This is frequently used in liberating hydrogen sulphide, carbon dioxide, and hydrogen, and for other purposes, and need not be pure. Where the dilute acid is called for as a reagent it must be made by mixing one part of the pure strong acid with four parts of distilled water.

Acid, acetic. Mix one part of the pure "glacial" acid with four parts of water.

Ammonia water. The strong solution is seldom used in analysis. The solution usually employed is made by mixing one volume of the stronger ammonia water (containing about 28 per cent of the gas) with three volumes of distilled water.

The solution should be free from carbonic acid, as presence of this would interfere with several of the tests where it is employed.

Ammonium carbonate. Dissolve one part of the pure powdered crystals in five parts of dilute ammonia water.

Ammonium chloride. Dissolve one part of the pure salt in ten parts of water.

Ammonium molybdate. A solution of this salt is chiefly used as a test for phosphoric acid, and should be prepared in this way: Dissolve 3 Gm. of the crystals in 20 Cc. of water, and pour this solution in 20 Cc. of strong nitric acid. Warm the mixture to about 40° C. (not above), and allow to settle.

As the reagent does not keep well, it should be made only in small quantities.

Ammonium oxalate. Dissolve one part of the pure crystals in twenty parts of water.

Ammonium sulphide. Dilute a quantity of strong ammonia water with an equal volume of water. Take three-fifths of this mixture and saturate it with hydrogen sulphide. Then add the remaining two-fifths of the diluted ammonia water and mix thoroughly.

Barium chloride. Dissolve one part of the crystals in ten parts of water.

Calcium hydroxide. Slake pure white lime and pour over it a large excess of water. Allow to settle and throw away the clear liquid. Again add pure water, shake thoroughly, allow to settle as before, and decant the clear liquid into bottles for use. These bottles must be tightly stoppered. The portion rejected contains small amounts of impurities, possibly present in the lime.

Lead acetate. One part of the pure crystals to ten parts of water. It may be necessary to add a few drops of acetic acid to secure a clear solution.

Lead acetate, basic. To make a liter of solution of this reagent weigh out 170 Gm. of lead acetate and 120 Gm. of yellow lead oxide.

Dissolve the lead acetate in 800 Cc. of boiling, distilled water, in a glass or porcelain vessel. Then add the oxide of lead and boil for half an hour, occasionally adding enough hot distilled water to make up the loss by evaporation. Remove the heat, allow the liquid to cool, and add enough distilled water, previously boiled and cooled, to make the product measure 1,000 Cc. Finally, filter the liquid in a covered funnel.

Solution of basic lead acetate should be kept in well-stoppered bottles.

Magnesia mixture. Dissolve 100 Gm. of magnesium sulphate and 100 Gm. of ammonium chloride in 800 Cc. of water, and add 100 Cc. of strong ammonia water. Allow the mixture to stand twenty-four hours, and filter.

Mercuric chloride. One part of the pure crystals to twenty parts of water.

Millon's reagent. Dissolve one part of mercury in two parts of strong nitric acid, by aid of heat finally, and after cooling dilute the solution with twice its volume of water.

Potassium bichromate. Dissolve one part of the pure crystals in ten parts of water. The dry powdered crystals are also used.

Potassium chromate. Dissolve one part of the crystals, free from chlorine, in ten parts of water.

Potassium ferrocyanide. Dissolve one part of the crystals in twenty parts of water.

Potassium hydroxide. Several solutions are employed in analytical chemistry. For most purposes one containing ten per cent of the "stick" hydroxide is sufficient.

Sodium hydroxide. Dissolve one part of the best "stick" hydroxide in ten parts of water, allow to settle, and decant the clear solution.

This solution acts on glass bottles, and soon deposits a sediment. Hence, a great deal of it should not be made at one time. The glass stoppers of bottles containing sodium hydroxide, and many other substances, should be covered with a thin layer of paraffine, which prevents their sticking fast.

Sodium hypochlorite. The "chlorinated soda" of the U. S. P. is to be used here, and the solution may be made in this manner: Weigh out 75 Gm. of good commercial "chloride of lime" and rub it up with 200 Cc. of water to a thin cream. Allow this to settle and pour the liquid through a filter. Stir the residue with a second 200 Cc. of water, pour the whole on the filter and wash the insoluble residue with 100 Cc. of water, allowing this to mix with the 400 Cc. Now dissolve 150 Gm. of sodium carbonate (crystals) in 800 Cc. of hot water and pour this into the other solution. Warm the mixed solution and stir it well. Pour the mixture on a filter and when the liquid has run through pour on water enough to bring the filtrate up to 1000 Cc. The solution so obtained should be kept in the dark.

Special Reagents.

Solutions for Sugar Tests.

Fehling solution. This has been referred to at length in Chapters II. and XII., and its preparation will now be given. In presence of alkali copper solutions are reduced by dextrose approximately according to this proportion:

$$5(CuSO_4 \cdot 5H_2O) : C_6H_{12}O_6.$$
$$1244.00 \quad : 179.58.$$

from which it follows that 34.64 Mg. of the crystallized sulphate in solution oxidizes 5 Mg. of dextrose in solution.

The Fehling solution proper consists of a mixture of copper sulphate, an alkali and a tartrate. Investigation has shown that as alkali sodium hydroxide is preferable and that Rochelle salt is the best tartrate for the purpose. It has also been found that the best results are obtained if the copper and tartrate are mixed just before needed for use. Therefore, prepare solutions separately as follows:

1. Dissolve 69.28 Gm. of pure, recrystallized copper sulphate in distilled water to make 1 liter of solution. Much of the copper sulphate sold by druggists contains ferrous sulphate and is not suitable for the purpose.

2. Dissolve 100 Gm. of sodium hydroxide in sticks in 500 Cc. of water, heat to boiling and add gradually 350 Gm. of pure recrystallized Rochelle salt. Stir until all is dissolved. Allow the solution to stand 24 hours in a covered vessel, then filter through asbestos into a liter flask and add water enough to make the solution 1 liter. The sodium hydroxide for this purpose should be the grade designated "precipitated by alcohol" and the Rochelle salt should be practically pure.

3. By mixing equal volumes of these two solutions the Fehling liquid is prepared, containing 34.64 Mg. of the copper sulphate in each cubic centimeter. This mixture is made when required for use.

The Loewe solution. According to Loewe copper solutions prepared with glycerol are more stable than are those with a tartrate, while the oxidizing power of the copper hydroxide is practically the same. For quantitative tests Loewe prepares a solution by mixing

    Crystallized copper sulphate............ 34.64 Gm.
    Pure glycerol... .... 26.00 Gm.
    Sodium hydroxide solution, 1.84 Sp. Gr. 70.00 Cc.

with a small amount of water and heating to dissolve, after which the solution is diluted to 1 liter.

When used as a qualitative test this solution is diluted by adding glycerol. In place of the crystallized copper sulphate Loewe recommends to weigh out the corresponding amount of precipitated, washed and dried copper hydroxide, which can be kept indefinitely in perfectly stable form.

Pavy's solution. The use of the Pavy liquid as a sugar test was explained in Chapter XII. It is prepared as follows:

Dissolve 34.64 Gm. of crystallized copper sulphate, 170 Gm. of Rochelle salt and 170 Gm. of good stick potassium hydroxide in distilled water to make 1 liter. Mix 120 Cc. of this solution with 400 Cc. of strong ammonia water, Sp. Gr. 0.88, and dilute with distilled water to 1 liter. The oxidizing power of this solution is assumed to be just one-tenth of that of the Fehling solution, which would follow if the reaction takes place in the proportion, $6\ (CuSO_4.5H_2O) : C_6H_{12}O_6$.

Loewe-Pavy solution, as recommended by Dr. Purdy.
This is made by dissolving

    Crystallized copper sulphate.............. 4.74 Gm.
    Potassium hydroxide..................... 23.50 ''
    Ammonia water, Sp. Gr. 0.90............ 450.00 Cc.
    Glycerol .... 38.00 ''

in enough water to make 1000 Cc.

Dissolve the copper sulphate and glycerol in 200 Cc. of the water. In another 200 Cc. dissolve the alkali. Mix the two solutions, cool, add the ammonia and make up to the 1000 Cc. Pure chemicals must be employed

According to Dr. Purdy 35 Cc. of this solution oxidizes 20 Mg. of dextrose.

Schmiedeberg's solution. Dissolve 34.64 Gm. of pure crystallized copper sulphate in 200 Cc. of water and 16 Gm. of mannitol in 100 Cc. of water. Mix the two solutions and add 480 Cc. of sodium hydroxide solution, having a specific gravity of 1.145. Dilute to 1 liter. This solution is assumed to have the oxidizing power of the Fehling solution.

Knapp's solution. This is made by dissolving 10 Gm. of dry mercuric cyanide in 600 Cc. of distilled water. To this solution is added 100 Cc. of a solution of sodium hydroxide, having a specific gravity of 1.145 and the mixture is diluted to 1 liter.

Sachsse's solution. Dissolve 18 Gm. of pure mercuric iodide and

25 Gm. of potassium iodide in water. Add a solution of 80 Gm. of good potassium hydroxide and dilute to 1 liter. These solutions are nearly equal in oxidizing power. It has been shown that 50 Cc. of Sachsse's solution is reduced by 168 Mg. of dextrose.

Solutions for Water Tests.

Nessler's reagent. Dissolve 13 Gm. of mercuric chloride and 35 Gm. of potassium iodide in 600 Cc. of water by aid of heat. When all is dissolved add a little more mercuric chloride until a trifling red precipitate remains. Dissolve 100 Gm. of potassium hydroxide in 200 Cc. of water, and when cold add this solution to the other. Allow the mixture to stand twenty-four hours and pour the clear solution from the brownish precipitate. Sensitize this liquid poured off by adding, a drop at the time, solution of mercuric chloride as long as a clear solution is left on shaking.

Alkaline permanganate solution. Dissolve 8 Gm. of potassium permanganate crystals in 800 Cc. of water, add 200 Gm. of potassium hydroxide in sticks and water to make 1200 Cc. Boil this solution down to 1000 Cc. in a porcelain dish, cool and pour into a bottle which close with a paraffined glass stopper.

Other solutions used in water tests are described in Chapter VIII.

Indicators.

The indicators most commonly employed in the titration of acids and alkalies are aqueous or alcoholic solutions of *litmus, cochineal, phenol phthalein, methyl orange* and *rosolic acid*. In addition to these certain others are employed for special purposes, and among these there may be mentioned, *congo red, benzopurpurin, methyl violet* and the *tropæolins*. A few explanations will be given on the preparation of the first.

Litmus. Crude litmus comes in commerce in the form of small blue cubes. These are powdered and extracted by hot water. The aqueous solution is concentrated and acidified with acetic acid, after which it is evaporated to a paste. Treat this with an excess of 85 per cent alcohol which dissolves foreign matters but leaves the true color. Throw the mixture on a filter and wash the residue with strong alcohol. Then dissolve the precipitated color on the filter by means of hot water and keep this aqueous solution for use in a bottle loosely stoppered as access of air is necessary for its preservation.

A neutral litmus solution is used for certain purposes and may be prepared by dividing the aqueous solution just described into two portions, one of which is rendered faintly acid by nitric acid, while the other is made alkaline by potassium hydroxide, added in drops of very dilute solution. On mixing these two liquids the product will be found practically neutral.

Cochineal. A solution is made by extracting 1 part of the crushed cochineal with 10 parts of weak alcohol. This indicator is valuable in titrating in presence of carbonic acid or ammonia.

Phenol phthalein Dissolve 1 part of the pure commercial product in 200 parts of 50 per cent alcohol. This indicator cannot be well used with ammonia or in presence of free carbonic acid.

Methyl orange. Dissolve 1 part of the color in 1,000 parts of distilled water. Although this solution is very weak a *single drop* is sufficient for an ordinary titration; with more the change of color is less characteristic or sharp The indicator is valuable in the titration of carbonates or ammoniacal liquids. Carbonic acid does not act on it.

Rosolic acid. Dissolve 1 part in 500 parts of 50 per cent alcohol. This is a sensitive indicator for the mineral acids.

Tables of Weights and Measures.

The Metric System.

1 Meter	= 100 Centimeters (Cm.)	= 1,000 Millimeters (Mm.)
	1 Liter = 1,000 Cubic Centimeters (Cc.)	
	1 Kilogram = 1,000 Grams (Gm.)	
	1 Gram = 1,000 Milligrams (Mg.)	

American Weights and Measures.

	1 Gallon	= 8 Pints.
	1 Pint	= 16 Fluidounces.
	1 Fluidounce	= 8 Fluidrachms.
	1 Fluidrachm	= 60 Minims.
1 Avoirdupois pound		= 16 Avoirdupois ounces.
1 Avoirdupois ounce		= 437½ Grains.
1 Apothecaries' ounce		= 8 Drachms.
1 Apothecaries' drachm		= 60 Grains.

Equivalents.

1 Meter	= 39.370	Inches.
1 Liter	= 33.815	U. S. Fluidounces.
1 Liter	= 35.219	Imp. Fluidounces.
1 Cubic Centimeter	= 16.231	U. S. Minims.
1 Kilogram	= 32.151	U. S. Apoth. ounces.
1 Kilogram	= 35.274	Avoirdupois ounces.
1 Gram	= 15.432	Grains.
1 U. S. fluidounce	= 29.57	Cc.
1 Imperial fluidounce	= 28.39	Cc.
1 U. S. Apoth. ounce	= 31.103	Gm.
1 Avoird. ounce	= 28.349	Gm.
1 Grain	= 64.798	Milligrams.

1 U. S. Fluidounce of Water weighs 0.95 U. S. Apoth. ounce.
96 U. S. Fluidounces of Water weigh 100 Avoirdupois ounces.
1 Imperial Fluidounce of Water weighs 1 Avoirdupois ounce.

Determination of the Specific Gravity of Liquids.

The specific gravity of liquids may be found by several methods, most accurately by means of the pycnometer or small weighing bottle.

For approximate tests specific gravity bulbs, illustrated by the urinometer already described, are very commonly used.

A very convenient and accurate apparatus for these determinations is the Mohr-Westphal balance, shown in the figure below. From the outer end of the beam is suspended a small weight, A, usually in the form of a thermometer, which hung in the air, holds the beam in equilib-

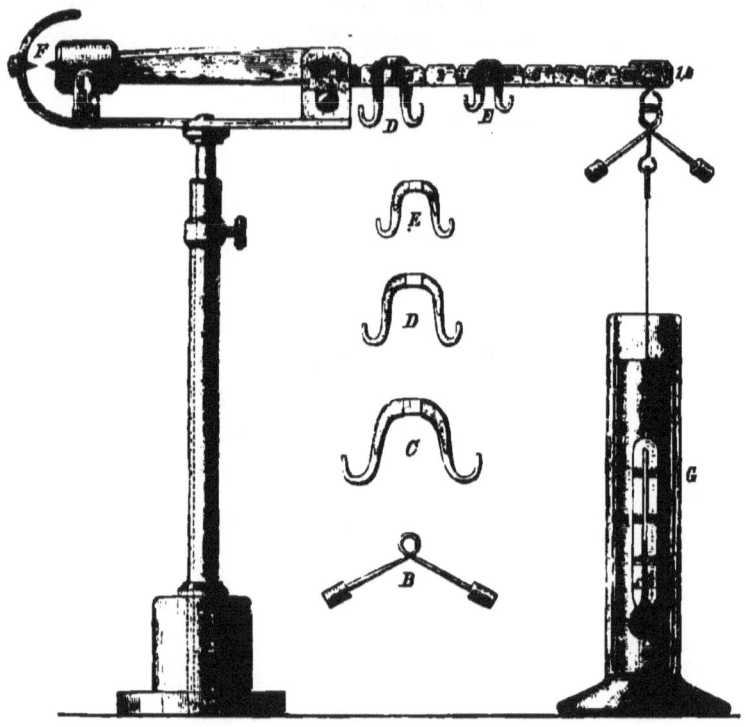

FIG. 69.

rium. If the weight is immersed in pure water in the jar G, a counterpoise, B, must be hung on the end of the beam to restore the equilibrium. This weight B, when hung on the beam, is therefore able to hold the latter in horizontal position when the liquid below has unit density. The beam is decimally divided and other specific gravities are found by the use of the riders shown. C has the same weight as B, $D = \frac{1}{10} C$, and $E = \frac{1}{10} D$. With the weights hung as in the figure the specific gravity of the liquid in the jar is 1.025. The apparatus gives a correct result for some definite temperature only, usually 15° C.

Approximate Specific Gravity Tables.

Hydrochloric Acid.		Nitric Acid.	
Sp. Gr. $\frac{15°}{4°}$	Parts of HCl by weight in 100 parts.	Sp. Gr. $\frac{15°}{4°}$	Parts of HNO_3 by weight in 100 parts.
1.000	0.16	1.00	0.10
1.005	1.15	1.01	1.90
1.010	2.14	1.02	3.70
1.015	3.12	1.03	5.50
1.020	4.13	1.04	7.26
1.025	5.15	1.05	8.99
1.030	6.15	1.06	10.68
1.035	7.15	1.07	12.33
1.040	8.16	1.08	13.95
1.045	9.16	1.09	15.53
1.050	10.17	1.10	17.11
1.055	11.18	1.11	18.69
1.060	12.19	1.12	20.23
1.065	13.19	1.13	21.77
1.070	14.17	1.14	23.31
1.075	15.16	1.15	24.84
1.080	16.15	1.16	26.36
1.085	17.13	1.17	27.88
1.090	18.11	1.18	29.38
1.095	19.06	1.19	30.88
1.100	20.01	1.20	32.36
1.105	20.97	1.21	33.82
1.110	21.92	1.22	35.28
1.115	22.86	1.23	36.78
1.120	23.82	1.24	38.29
1.125	24.78	1.25	39.82
1.130	25.75	1.26	41.34
1.135	26.70	1.27	42.87
1.140	27.66	1.28	44.41
1.145	28.61	1.29	45.95
1.150	29.57	1.30	47.49
1.155	30.55	1.31	49.07
1.160	31.52	1.32	50.71
1.165	32.49	1.33	52.37
1.170	33.46	1.34	54.07
1.175	34.42	1.35	55.79
1.180	35.39	1.36	57.57
1.185	36.31	1.37	59.39
1.190	37.23	1.38	61.27
1.195	38.16	1.39	63.23
1.200	39.11	1.40	65.30
		1.41	67.50
		1.42	69.80
		1.43	72.17
		1.44	74.68
		1.45	77.28

SULPHURIC ACID.

Sp. Gr. $\frac{15°}{4°}$	Parts of H_2SO_4 by wt. in 100 parts.	Sp. Gr. $\frac{15°}{4°}$	Parts of H_2SO_4 by wt. in 100 parts.
1.00	0.09	1.46	55.97
1.01	1.57	1.47	56.90
1.02	3.03	1.48	57.81
1.03	4.49	1.49	58.74
1.04	5.96	1.50	59.70
1.05	7.37	1.51	60.65
1.06	8.77	1.52	61.59
1.07	10.19	1.53	62.53
1.08	11.60	1.54	63.43
1.09	12.99	1.55	64.26
1.10	14.35	1.56	65.01
1.11	15.71	1.57	65.90
1.12	17.01	1.58	66.71
1.13	18.31	1.59	67.59
1.14	19.61	1.60	68.51
1.15	20.91	1.61	69.43
1.16	22.19	1.62	70.32
1.17	23.47	1.63	71.16
1.18	24.76	1.64	71.99
1.19	26.04	1.65	72.82
1.20	27.32	1.66	73.64
1.21	28.58	1.67	74.51
1.22	29.84	1.68	75.42
1.23	31.11	1.69	76.30
1.24	32.28	1.70	77.17
1.25	33.43	1.71	78.04
1.26	34.57	1.72	78.92
1.27	35.70	1.73	79.80
1.28	36.87	1.74	80.68
1.29	38.03	1.75	81.56
1.30	39.19	1.76	82.44
1.31	40.35	1.77	83.32
1.32	41.50	1.78	84.50
1.33	42.66	1.79	85.70
1.34	43.74	1.80	86.90
1.35	44.82	1.81	88.30
1.36	45.88	1.82	90.05
1.37	46.94	1.83	92.10
1.38	48.00	1.8339	93.00
1.39	49.06	1.8372	94.00
1.40	50.11	1.8390	95.00
1.41	51.15	1.8406	96.00
1.42	52.15	1.8410	97.00
1.43	53.11	1.8412	98.00
1.44	54.07	1.8403	99.00
1.45	55.03	1.8384	100.00

APPENDIX.

	Potassium Hydroxide.				Sodium Hydroxide.		
Sp. Gr. 15°.	Parts of KOH in 100 parts by weight.	Sp. Gr. 15°.	Parts of KOH in 100 parts by weight.	Sp. Gr. 15°.	Parts of NaOH in 100 parts by weight.	Sp. Gr. 15°.	Parts of NaOH in 100 parts by weight.
1.009	1	1.361	36	1.012	1	1.395	36
1.017	2	1.374	37	1.023	2	1.405	37
1.025	3	1.387	38	1.035	3	1.415	38
1.033	4	1.400	39	1.046	4	1.426	39
1.041	5	1.412	40	1.058	5	1.437	40
1.049	6	1.425	41	1.070	6	1.447	41
1.058	7	1.438	42	1.081	7	1.457	42
1.065	8	1.450	43	1.092	8	1.468	43
1.074	9	1.462	44	1.103	9	1.478	44
1.083	10	1.475	45	1.115	10	1.488	45
1.092	11	1.488	46	1.126	11	1.499	46
1.101	12	1.499	47	1.137	12	1.509	47
1.110	13	1.511	48	1.148	13	1.519	48
1.119	14	1.525	49	1.159	14	1.529	49
1.128	15	1.539	50	1.170	15	1.540	50
1.137	16	1.552	51	1.181	16	1.550	51
1.146	17	1.565	52	1.192	17	1.560	52
1.155	18	1.578	53	1.202	18	1.570	53
1.166	19	1.590	54	1.213	19	1.580	54
1.177	20	1.604	55	1.225	20	1.591	55
1.188	21	1.618	56	1.236	21	1.601	56
1.198	22	1.630	57	1.247	22	1.611	57
1.209	23	1.642	58	1.258	23	1.622	58
1.220	24	1.655	59	1.269	24	1.633	59
1.230	25	1.667	60	1.279	25	1.643	60
1.241	26	1.681	61	1.290	26	1.654	61
1.252	27	1.695	62	1.300	27	1.664	62
1.264	28	1.705	63	1.310	28	1.674	63
1.276	29	1.718	64	1.321	29	1.684	64
1.288	30	1.729	65	1.332	30	1.695	65
1.300	31	1.740	66	1.343	31	1.705	66
1.311	32	1.754	67	1.353	32	1.715	67
1.324	33	1.768	68	1.363	33	1.726	68
1.336	34	1.780	69	1.374	34	1.737	69
1.349	35	1.790	70	1.384	35	1.748	70

AMMONIA WATER.

Sp. Gr. $\frac{15°}{4°}$	Parts of NH_3 in 100 parts by weight.	Sp. Gr. $\frac{15°}{4°}$	Parts of NH_3 in 100 parts by weight.
1.000	0.00	0.940	15.63
0.998	0.45	0.938	16.22
0.996	0.91	0.936	16.82
0.994	1.37	0.934	17.42
0.992	1.84	0.932	18.03
0.990	2.31	0.930	18.64
0.988	2.80	0.928	19.25
0.986	3.30	0.926	19.87
0.984	3.80	0.9256	20.00
0.982	4.30	0.924	20.49
0.980	4.80	0.922	21.12
0.9792	5.00	0.920	21.75
0.978	5.30	0.918	22.39
0.076	5.80	0.916	23.03
0.974	6.30	0.914	23.68
0.972	6.80	0.912	24.33
0.970	7.31	0.910	24.99
0.968	7.82	0.9099	25.00
0.966	8.33	0.908	25.65
0.964	8.84	0.906	26.31
0.962	9.35	0.904	26.98
0.960	9.91	0.902	27.65
0.9597	10.00	0.900	28.33
0.958	10.47	0.898	29.01
0.956	11.03	0.896	29.69
0.954	11.60	0.8951	30.00
0.952	12.17	0.894	30.37
0.950	12.74	0.892	31.05
0.948	13.31	0.890	31.75
0.946	13.88	0.888	32.50
0.944	14.46	0.886	33.25
0.9421	15.00	0.884	34.10
0.942	15.04	0.882	34.95

Alcohol.

Per cent by volume.	Sp. Gr. $\frac{15.56°}{15.56°}$	Per cent by volume.	Sp. Gr. $\frac{15.56°}{15.56°}$	Per cent by volume.	Sp. Gr. $\frac{15.56°}{15.56°}$
0	1.00000	34	0.96043	68	0.89499
1	0.99847	35	0.95910	69	256
2	699	36	778	70	010
3	555	37	632	71	0.88762
4	415	38	487	72	511
5	279	39	338	73	257
6	147	40	185	74	000
7	019	41	029	75	0.87740
8	0.98895	42	0.94868	76	477
9	774	43	704	77	211
10	657	44	536	78	0.86943
11	543	45	364	79	670
12	432	46	188	80	395
13	324	47	008	81	116
14	218	48	0.93824	82	0.85833
15	114	49	636	83	547
16	011	50	445	84	256
17	0.97909	51	250	85	0.84961
18	808	52	052	86	660
19	708	53	0.92850	87	355
20	608	54	646	88	044
21	507	55	439	89	0.83726
22	406	56	229	90	400
23	304	57	015	91	065
24	201	58	0.91799	92	0.82721
25	097	59	580	93	365
26	0.96991	60	358	94	0.81997
27	883	61	134	95	616
28	772	62	0.90907	96	217
29	658	63	678	97	0.80800
30	541	64	447	98	359
31	421	65	214	99	0.79891
32	298	66	0.89978	100	0.79391
33	172	67	740		

Volume and Specific Gravity of Water of Different Temperatures.

(Calculations of Volkmann from results of Kopp, Hagen, Matthiesen, Jolly and Pierre.)

Temperature C.	Specific gravity, or weight of 1 Cc. of water in vacuo in grams.	Differences for 1°, Sp. Gr. and volume.	Volumes of one gram of water in Cc.
0	0.99988	0.00005	1.00012
1	0.99993	.00004	1.00007
2	0.99997	.00002	1.00003
3	0.99999	.00001	1.00001
4	1.00000	.00001	1.00000
5	0.99999	.00002	1.00001
6	0.99997	.00004	1.00003
7	0.99993	.00005	1.00007
8	0.99988	.00006	1.00012
9	0.99982	.00008	1.00018
10	0.99974	.00009	1.00026
11	0.99965	.00011	1.00035
12	0.99954	.00011	1.00046
13	0.99943	.00013	1.00057
14	0.99930	.00015	1.00070
15	0.99915	.00015	1.00085
16	0.99900	.00016	1.00100
17	0.99884	.00018	1.00116
18	0.99866	.00019	1.00134
19	0.99847	.00020	1.00153
20	0.99827	.00021	1.00173
21	0.99807	.00022	1.00194
22	0.99785	.00023	1.00216
23	0.99762	.00023	1.00238
24	0.99739	.00025	1.00262
25	0.99714	.00027	1.00287
30	0.99577		1.00425

Specific Gravity and Volume of Mercury for Certain Temperatures.

Temp. C.	Sp. Gr. or wt. of 1 Cc. in grams.	Log.	Vol. of 1 Gm. in Cc.	Log.
0	13.5953	1.13339	0.073555	− 2.86661
4	13.5854	1.13307	0.073608	2.86693
5	13.5830	1.13299	0.073622	2.86701
10	13.5707	1.13260	0.073688	2.86740
15	13.5584	1.13221	0.073755	2.86779
20	13.5461	1.13182	0.073822	2.86818
25	13.5339	1.13142	0.073888	2.86858
30	13.5217	1.13103	0.073955	2.86897

Tension of Aqueous Vapor at Different Temperatures, Expressed in Millimeters of Mercury Pressure Equivalent.

Temp. C.	Hg. in Mm.	Temp. C.	Hg. in Mm.	Temp. C.	Hg. in Mm.
0	4.57	11	9.77	22	19.63
1	4.91	12	10.43	23	20.86
2	5.27	13	11.14	24	22.15
3	5.66	14	11.88	25	23.52
4	6.07	15	12.67	26	24.96
5	6.51	16	13.51	27	26.47
6	6.97	17	14.40	28	28.06
7	7.47	18	15.33	29	29.74
8	7.99	19	16.32	30	31.51
9	8.55	20	17.36	31	33.37
10	9.14	21	18.47	32	35.32

The Relation Between the Barometric Pressure and Boiling Point of Water.

(FOR CORRECTION OF THERMOMETERS.)

Barometric height in Mm.	Boiling Point C.	Barometric height in Mm.	Boiling Point C.	Barometric height in Mm.	Boiling Point C.
710	98.11	731	98.92	752	99.70
711	98.15	732	98.96	753	99.74
712	98.19	733	98.99	754	99.78
713	98.23	734	99.03	755	99.82
714	98.27	735	99.07	756	99.85
715	98.30	736	99.11	757	99.89
716	98.34	737	99.14	758	99.92
717	98.38	738	99.18	759	99.96
718	98.42	739	99.22	760	100.00
719	98.46	740	99.26	761	100.03
720	98.50	741	99.30	762	100.07
721	98.54	742	99.33	763	100.11
722	98.57	743	99.37	764	100.14
723	98.61	744	99.41	765	100.18
724	98.65	745	99.44	766	100.22
725	98.69	746	99.48	767	100.26
726	98.73	747	99.52	768	100.29
727	98.77	748	99.55	769	100.33
728	98.80	749	99.59	770	100.36
729	98.84	750	99.63	771	100.40
730	98.88	751	99.67	772	100.44

Formula for Correction of Thermometer Reading on Account of Projecting Thread.

$T =$ the corrected temperature.
$x =$ the thermometer reading observed.
$x' =$ the mean temperature of projecting thread.
$p =$ the length of the projecting thread in degrees.
0.000143 is a factor determined by Thorpe empirically.
$T = x + 0.000143\, p\, (x - x')$.

For example: If 150° is the observed reading, and 70 is the number of degrees outside the hot vapor or liquid in which the thermometer is suspended, while the mean temperature of this part of the thread is 30° (as shown by a second small thermometer hanging against the first) the correction, $0.000143\, p\, (x - x')$, is 1.20 and the corrected reading 151.2.°

Table Showing the Average Composition of Several Vegetable and Animal Food Stuffs.

	Water.	Pro-teids.	Fat.	Carbo-hydrate	Cellu-lose.	Ash.
Wheat	13.65	12.35	1.75	67.91	2.53	1.81
Rye	15.06	11.52	1.79	67.81	2.01	1.81
Barley	13.77	11.14	2.16	64.93	5.31	2.69
Oats	12.37	10.41	5.23	57.78	11.19	3.02
Corn	13.12	9.85	4.62	68.41	2.49	1.51
Buckwheat, with hulls	11.93	10.30	2.81	55.81	16.43	2.72
Rice	13.11	7.85	0.88	76.52	0.63	1.01
Peas, ripe	14.99	22.85	1.79	52.36	5.43	2.58
Beans, ripe	14.76	24.27	1.61	49.01	7.09	3.26
Peas, green	78.44	6.35	0.53	12.00	1.87	0.81
Beans, green	84.07	5.43	0.33	7.35	2.08	0.74
Potatoes	75.48	1.95	0.15	20.69	0.75	0.98
Beets	87.71	1.09	0.11	9.26	0.98	0.95
Fat steer	55.42	17.19	26.38			1.08
Medium steer	72.25	20.91	5.19			1.17
Fat sheep	47.91	14.80	36.39			0.85
Medium sheep	75.99	17.11	5.77			1.33
Fat hog	47.40	14.54	37.34			0.72
Lean hog	72.57	20.25	6.81			1.10
Horse flesh	74.27	21.71	2.55			1.01

The values given for the animal products represent the whole animal.

List of Apparatus Required in Performing the Experiments of This Book.

The following list embraces the important apparatus called for in the experiments of the Chemical Physiology and Urine Analysis. When thought necessary some of the pieces may be used in common by several students. The apparatus can be purchased by students at the attached prices, kindly furnished the author by Messrs. E. H. Sargent & Co., of Chicago:

2 beakers, 100 Cc., each	10
2 beakers, 250 Cc., each	18
2 beakers, 400 Cc., each	25
2 flasks, 100 Cc., each	15
2 flasks, 250 Cc., each	20
2 flasks, 500 Cc., each	25
1 rack and 10 6-inch test-tubes	75
1 Bunsen burner and rubber tube	75
1 porcelain dish, 6 inches	45
2 porcelain dishes, 3 inches, each	20
1 test-tube holder, wooden	10
1 test-tube brush, sponge end	10
1 50 Cc. burette, g. s.	2.00
1 50 Cc. pipette	40
1 25 Cc. pipette	35
1 5 Cc. pipette	15
1 100 Cc. graduate	50
1 100° thermometer	85
1 funnel tube	15
2 funnels, 70 Mm., each	15
1 horn spatula, 6 inches	15
1 retort stand with 3 rings	70
1 6-inch copper water-bath with rings	1.50
1 piece of wire gauze, 4 inches square	10

-In addition to these items the student will need an assortment of corks, glass tubing, rubber connections, files and a few other small articles.

INDEX.

Abnormal coloring matters	246
Absorption analysis	156–165
Acetic acid fermentation	55
Aceto-acetic acid	240
Acetone in urine	238
Acetonuria	238
Acid albumin	74
Air tests	142
Albumins	67
classification of	70
composition of	67
in urine	187
reactions for	68
Albumose	81 and 204
amount of	206
tests for	83
Alcohol tables	355
Alcoholic fermentation	54
Alkali albumin	73
Alkapton	251
Almen's test	250
Alpha-naphthol test	224
Ammonia water	354
Amorphous phosphates	331
urates	323
Amount of albumin	198
albumose	206
phosphates	285
sugar	227
uric acid	257
Amyloid substance	89
Analysis of calculi	332–339
of meat extract	169–173
by the spectroscope	156–165
Animal starch	57
Antialbumose	82
Antipeptone	82
Aqueous vapor tension	357
Atomic weights	343
Bacilli	318
Baryta solution	267
Beef extract	127–131
Bile	117
Bile acid	117 and 252
action on fats	119
pigments	118
colors in urine	246
Bilirubin	117
Biliverdin	117
Bismuth test	220
Biuret test	69 and 208
Blood	91–101
coagulation of	94
coloring matters	248
composition of	91 and 92
corpuscles	108 and 302
corpuscle counters	109 and 110
spectra	105 and 106
tests	92
Boettger's bismuth test	220
Bone constituents	111
Bright's disease	188
British gum	29
Bruecke's solution	221
Burettes	11
Burette apparatus	277
Butyric fermentation	56
Calcium carbonate	328
oxalate	328
phosphate	331
sulphate	328

Calculi	332
Cancer tissue	316
Carbohydrates	19-59
Casein	75
Casts	309-314
Centrifugal machine	300
Chlorides in urine	289
Chlorides in water	135
Cholesterin	327
Chrysophanic acid	250
Classification of albumins	70
of sediments	302
Coagulated blood	94
proteids	80
Coagulation test	189
Colloids	16
Coloring matter in urine	242
Color of urine	185
Composition of bone	111
of milk	121
Corpuscle counters	109
Crystallin	78
Crystalloids	16
Cystin	326
Determination of albumin	198
of chlorides	289
of nitrogen	165
of phosphates	285
of sugars	35
of urea	265
of uric acid	254
Derived albumins	73
Dextrin	29
Dextrose	30
tests for	31-33
Diabetes mellitus	182
Diacetic acid	240
Dialysis	15
Dialyzer	17
Diffusion of peptones	89
Digestion of proteids	82
Doremus' apparatus	281
Double iodide test	194
Egg albumin	71
Emulsions	62
Epithelium	306
Esbach's albuminometer	200
Expired air	141
Extract of meat	127
of pancreas	28
Fats	59-66
crystallization of	63
crystals	64 and 65
in milk	122
origin of	59
saponification	60
Fatty acids	61
Fehling's solution	35
test	217
Fermentation of sugar	54
test	225
Ferrocyanide test	196
Fibrin	80
Fibrinogin	79
Fleischl's hæmometer	100
Flour	131
Food stuffs	359
Foreign matters in urine	332
Fungi	316
Gastric juice	114
tests of	114
Globin	80
Globulins	77
Glycerol	62
Glycocholic acid	148
Glycocoll	146
Glycogen	57
Gmelin's test	246
Gowers' hæmoglobinometer	98

Gravimetric method	199	Maltose	36
Guaiacum test	92	Meal	131
Gunning's method	166	Measuring apparatus	11-12
		Meat extract	127 and 169
Hæmometer	100	Mercuric nitrate solution	267
Haeser's coeffiicient	181	Mercuric-potassium iodide test	194
Hæmin crystals	93	Mercury, specific gravity	357
Haycraft's method	259	Methæmoglobin	105
Heat test	189	Micrococcus ureæ	317
Heller's test	247 and 249	Milk	121
Hæmoglobin	94	peptonization of	126
amount of	97-101	sugar	34
Hæmoglobinometer	98	Millon's reagent	69
Hemipeptone	82	Myosin	79
Hippuric acid	263	Moliscb's test	33
Hofmeister's test	208	Moore's test	215
Huefner's apparatus	274	Mucin	211-212
Hydrochloric acid	351	bands	308-309
		Mucus corpuscles	304-305
Indican	244	Murexid test	255
Indicators	348 and 349		
Inosite in urine	237	Native albumins	79
Introduction	1	Nicol's prism	40-44
Invert sugar	36	Nitric acid	351
		Nitric acid test	191
Kjeldahl process	165-169	Nitrates and nitrites	139
Knapp's solution	232	Nitrogen, determination of	165
		Normal colors	242
Lactic fermentation	56		
Lactose in urine	237	Oder of urine	184
Lævulose in urine	236	Organic compounds	3
Lardacein	89	classification of	5
Legal's test	239	Organized sediments	301-302
Leucin	82	Outline of urine tests	178
and tyrosin	149 and 324-326	Oxidation tests	137
preparation of	149 and 140		
Liberation of nitrogen	273	Pancreatic digestion	87
Lieben's test	239	Paraglobulin	78
Lieberkuehn's jelly	74	Pavy's solution	230
Liebig's method	265	Pepsin, tests of	84
Loewe's test	220	Peptic digestion	83
Magnesium phosphate	331	Peptones	81 and 206

Peptones, diffusion of......... 89
    tests for................. 83
Peptonized milk.............. 126
Peptonuria................... 206
Phenols in urine.............. 251
Phenylhydrazine test.. 223
Phosphates................... 329
    in urine................ 282
Picric acid test............... 197
Pigments of bile.............. 118
Polarimetry................... 233
Polariscopes.................. 39
Polarization of light...... 40
    methods................39-53
Potassium hydroxide.......... 353
Preliminary tests............. 177
Preparation of urea........... 143
Preservation of sediments..... 299
Proteids.....................67-90
    in milk................... 124
    separation of... 160
Pus corpuscles............... 304

Reaction of blood............. 98
    of urine......... 183
Reagents............. . ..344-348
Rennet....................... 125

Saccharoses.................. 33
Sachsse's solution... 232
Salicylic acid................. 251
Saliva....................... 113
Salkowski-Ludwig method..... 256
Santonin..................... 250
Schmiedeberg's test..... 220
Sediment of urine............. 298
Serum albumin.........71 and 188
    globulin................. 202
Silver nitrate solution..291 and 294
Sodium chloride solution...... 292
Sodium hydroxide............ 353
Special problems............. 143

Specific gravity....... 17 and 349
    of urine................. 179
    tables................351-357
Specific rotation.............. 51
Spectra of blood.............. 105
Spectro-photometer........... 150
Spectro-photometry..... ...150 165
Spectroscope................. 101
Spermatozoa................. 315
Squibb's apparatus........... 279
Standard solutions........... 8
Starch.....................19-29
    and acids......... 22
    and malt extract.......... 26
    and pancreatic extract.... 27
    and saliva................ 25
    preparation......... 19
    properties...... 20
Struve's test................. 249
Stutzer's method.. 169
Sugars......................29-58
    determination of.......... 35
    in milk.................. 123
    in urine................. 213
    reactions................. 226
Sulphocyanate solution........ 294
Sulphuric acid................ 352
Syntonin.................... 77
Synthesis.................... 146

Tanret's solution............. 194
Tension of aqueous vapor...... 357
Test solutions.............344-348
Tests,
    Almen's........ 250
    alpha-naphthol........... 224
    bismuth................. 220
    biuret............69 and 208
    Boettger's............... 220
    coagulation.............. 189
    congo red................ 115

Tests,
   Donne's 305
   double iodide 194
   emerald green............ 115
   Esbach's.................. 200
   Fehling............31 and 217
   fermentation 225
   ferrocyanide 196
   Gmelin's..........118 and 246
   guaiacum................. 92
   Haycraft's 259
   Heller's ....247 and 249
   Hofmeister's.............. 208
   Knapp's 232
   Legal's................... 239
   Lieben's.................. 239
   Loewe's 220
   methyl violet............. 116
   Molisch's 33
   Moore's 215
   murexid 255
   Nessler 136
   oxidation 137
   Pavy's 230
   permanganate 138
   Pettenkofer's.............. 117
   phenyl hydrazine...33 and 223
   phloroglucin and vanillin.. 116
   picric acid 197
   polarization 39
   Sachsse's 232
   Schmiedeberg's 220
   specific gravity 179
   Struve's 249
   sulphate 203
   Tanret's 194
   Trommer's.........30 and 215
   tropæolin 116
   Trousseau's............... 247
   Weyl's 130
   xantho-proteic............ 69

Thermometers 13
   correction 358
Triple phosphate............ 330
Trommer's test............. 215
Trousseau's test 247
Tyrosin ....82 and 149
   preparation of....149 and 150

Unorganized sediment..301 and 321
Uranium solution............ 286
Urates.... 323
Urea....................... 263
   preparation of........143–145
   solution.................. 268
Uric acid.............245 and 321
   synthesis of.............. 146
Urinary calculi.............. 332
Urine, aceto-acetic acid in..... 240
   acetone in................ 238
   albumins in.............. 187
   albumose in.............. 204
   alkapton in............... 251
   analysis of,.............. 177
   bile acids in.............. 252
   biliary colors in.......... 246
   blood colors in........... 248
   blood corpuscles in........ 302
   calcium oxalate in........ 328
   cancer tissue in........... 316
   casts in.................. 309
   chlorides in.............. 289
   cholesterin in. 327
   chrysophanic acid in...... 250
   color 185
   colors in................. 242
   epithelium in............. 306
   fungi in.................. 316
   hippuric acid in.......... 262
   indican in................ 244
   lactose in................ 237
   lævulose in..... 236

Urine,
- leucin and tyrosin in....... 324
- mucin in.................. 211
- mucin bands in........... 308
- mucus and pus in......... 304
- oder..................... 184
- peptone in................ 206
- phenols in................ 251
- phosphates in......282 and 329
- reaction of............... 183
- salicylic acid in........... 251
- santonin in............... 250
- sediment of............... 298
- serum globulin in......... 202
- specific gravity of......... 179
- spermatozoa in........... 315
- sugar in.................. 213
- urea in................... 263
- uric acid in........254 and 321
- urobilin in................ 242

Urine,
- urohæmatin in............ 244
- urophain in.............. 244

Urinometer...... 180
Urobilin..................... 242
Urohæmatin 244
Urophain 244

Vitellin...................... 78
Volhard's method... 293
Volume method.............. 200

Water 134
- analysis of............135–144
- boiling point............. 358
- specific gravity of......... 358

Weights and measures......... 349
Weyl's reaction 130

Xanthoproteic reaction........ 69

Yeast........................ 132

www.ingramcontent.com/pod-product-compliance
Lightning Source LLC
Chambersburg PA
CBHW031421230426
43668CB00007B/385